ASE Test Preparation Series

Medium/Heavy Duty Truck Test

Heating, Ventilation, and Air Conditioning (Test T7)

4th Edition

THOMSON
DELMAR LEARNING™

Australia Canada Mexico Singapore Spain United Kingdom United States

Thomson Delmar Learning's ASE Test Preparation Series
Medium/Heavy Duty Truck Test for Heating, Ventilation, and Air Conditioning (Test T7), 4th Edition

Vice President, Technology Professional Business Unit:
Gregory L. Clayton

Product Development Manager:
Kristen L. Davis

Product Manager:
Kimberley Blakey

Editorial Assistant:
Vanessa Carlson

Director of Marketing:
Beth A. Lutz

Marketing Manager:
Brian McGrath

Marketing Coordinator:
Jennifer Stall

Production Director:
Patty Stephan

Production Manager:
Andrew Crouth

Content Project Manager:
Kara A. DiCaterino

Art Director:
Robert Plante

Cover Design:
Michael Egan

Printed in the United States of America
1 2 3 4 5 TW 10 09 08 07 06

For more information contact Thomson Delmar Learning
Executive Woods
5 Maxwell Drive, PO Box 8007,
Clifton Park, NY 12065-8007
Or find us on the World Wide Web at
www.delmarlearning.com

ISBN: 1-4180-4835-6

NOTICE TO THE READER

Publisher does not warrant or guarantee any of the products described herein or perform any independent analysis in connection with any of the product information contained herein. Publisher does not assume, and expressly disclaims, any obligation to obtain and include information other than that provided to it by the manufacturer.

The reader is expressly warned to consider and adopt all safety precautions that might be indicated by the activities herein and to avoid all potential hazards. By following the instructions contained herein, the reader willingly assumes all risks in connection with such instructions.

The publisher makes no representation or warranties of any kind, including but not limited to, the warranties of fitness for particular purpose or merchantability, nor are any such representations implied with respect to the material set forth herein, and the publisher takes no responsibility with respect to such material. The publisher shall not be liable for any special, consequential, or exemplary damages resulting, in whole or part, from the readers' use of, or reliance upon, this material.

Contents

Section 5 Sample Test for Practice

Section 6 Additional Test Questions for Practice

Section 7 Appendices

Preface

Delmar Learning is very pleased that you have chosen our ASE Test Preparation Series to prepare yourself for the ASE Medium/Heavy Truck Examination. These guides are available for all of the medium/heavy truck areas including T1-T8. These guides are designed to introduce you to the Task List for the test you are preparing to take, give you an understanding of what you are expected to be able to do in each task, and take you through sample test questions formatted in the same way the ASE tests are structured.

If you have a basic working knowledge of the discipline you are testing for, you will find Delmar Learning's ASE Test Preparation Series to be an excellent way to understand the "must know" items to pass the test. These books are not textbooks. Their objective is to prepare the technician who has the requisite experience and schooling to challenge ASE testing. It cannot replace the hands-on experience or the theoretical knowledge required by ASE to master vehicle repair technology. If you are unable to understand more than a few of the questions and their explanations in this book, it could be that you require either more shop-floor experience or further study. Some resources that can assist you with further study are listed on the rear cover of this book.

Each book begins with an item-by-item overview of the ASE Task List with explanations of the minimum knowledge you must possess to answer questions related to the task. Following that, there are two sets of sample questions followed by an answer key to each test and an explanation of the answers to each question. A few of the questions are not strictly ASE format but were included because they help teach a critical concept that will appear on the test. We suggest that you read the complete Task List Overview before taking the first sample test. After taking the first test, score yourself and read the explanation to any questions that you were not sure about, including the questions you answered correctly. Each test question has a reference back to the related task or tasks that it covers. This will help you to go back and read over any area of the task list that you are having trouble with. Once you are satisfied that you have all of your questions answered from the first sample test, take the additional tests and check them. If you pass these tests, you will be prepared to do well on the ASE test.

Our Commitment to Excellence

The 4th edition of Delmar Learning's ASE Test Preparation Series has been through a major revision with extensive updates to the ASE's task lists, test questions, and answers and explanations. Delmar Learning has sought out the best technicians in the country to help with the updating and revision of each of the books in the series.

To promote consistency throughout the series, a series advisor took on the task of reading, editing, and helping each of our experts give each book the highest level of accuracy possible.

Thank you for choosing Delmar Learning's ASE Test Preparation Series. All of the writers, editors, and Delmar Staff have worked very hard to make this series second to none. It is our objective to constantly improve our product at Delmar by responding to feedback.

If you have any questions concerning the books in this series, email us at: autoexpert@trainingbay.com.

1 The History and Purpose of ASE

ASE began as the National Institute for Automotive Service Excellence (NIASE). It was founded as a non-profit independent entity in 1972 by a group of industry leaders with the single goal of providing a means for consumers to distinguish between incompetent and competent technicians. It accomplishes this goal by testing and certification of repair and service professionals. From this beginning it has evolved to be known simply as ASE (Automotive Service Excellence) and today offers more than 40 certification exams in automotive, medium/heavy duty truck, collision, engine machinist, school bus, parts specialist, automobile service consultant, and other industry-related areas. At this time there are more than 400,000 professionals with current ASE certifications. These professionals are employed by new car and truck dealerships, independent garages, fleets, service stations, franchised service facilities, and more. ASE continues its mission by also providing information that helps consumers identify repair facilities that employ certified professionals through its Blue Seal of Excellence Recognition Program. Shops that have a minimum of 75 percent of their repair technicians ASE certified and meet other criteria can apply for and receive the Blue Seal of Excellence Recognition from ASE.

ASE recognized that educational programs serving the service and repair industry also needed a way to be recognized as having the faculty, facilities, and equipment to provide a quality education to students wanting to become service professionals. Through the combined efforts of ASE, industry, and education leaders, the non-profit National Automotive Technicians Education Foundation (NATEF) was created to evaluate and recognize training programs. Today more than 2,000 programs are ASE certified under the standards set by the service industry. ASE/NATEF also has a certification of industry (factory) training program known as CASE. CASE stands for Continuing Automotive Service Education and recognizes training provided by replacement parts manufacturers as well as vehicle manufacturers.

ASE certification testing is administered by the American College Testing (ACT). Strict standards of security and supervision at the test centers ensure that the technician who holds the certification earned it. Additionally, ASE certification also requires that the person passing the test be able to demonstrate that they have two years of work experience in the field before they can be certified. Test questions are developed by industry experts that are actually working in the field being tested. There is more detail on how the test is developed and administered in the next section. Paper-and-pencil tests are administered twice a year at over 700 locations in the United States. Computer-based testing is now also available with the benefit of instant test results at certain established test centers. The certification is valid for five years and can be recertified by retesting. So that consumers can recognize certified technicians, ASE issues a jacket patch, certificate, and wallet card to certified technicians and makes signs available to facilities that employ ASE certified technicians.

You can contact ASE at any of the following:

National Institute for Automotive Service Excellence
101 Blue Seal Drive S.E.
Suite 101
Leesburg, VA 20175
Telephone 703-669-6600
FAX 703-669-6123
www.ase.com

2 Take and Pass Every ASE Test

Participating in an Automotive Service Excellence (ASE) voluntary certification program gives you a chance to show your customers that you have the "know-how" needed to work on today's modern vehicles. The ASE certification tests allow you to compare your skills and knowledge to the automotive service industry's standards for each specialty area.

If you are the "average" automotive technician taking this test, you are in your mid-30s and have not attended school for about 15 years. That means you probably have not taken a test in many years. Some of you, on the other hand, have attended college or taken postsecondary education courses and may be more familiar with taking tests and with test-taking strategies. There is, however, a difference in the ASE test you are preparing to take and the educational tests you may be accustomed to.

How are the tests administered?

ASE test are administered at over 750 test sites in local communities. Paper-and-pencil tests are the type most widely available to technicians. Each tester is given a booklet containing questions with charts and diagrams where required. You can mark in this test booklet, but no information entered in the booklet is scored. Answers are recorded on a separate answer sheet. You will enter your answers, using a number 2 pencil only. ASE recommends you bring four sharpened number 2 pencils that have erasers. Answer choices are recorded by coloring in the blocks on the answer sheet. The answer sheets are scanned electronically and the answers tabulated. For test security, test booklets include randomly generated questions. Your answer key must be matched to the proper booklet so it is important to correctly enter the booklet serial number on the answer sheet. All instructions are printed on the test materials and should be followed carefully.

ASE has introduced Computer-based testing (CBT) at some locations. While the test content is the same for both testing methods the CBT tests have some unique requirements and advantages. It is strongly recommended that technicians considering the CBT tests go the ASE web page at www.ase.com and review the conditions and requirements for this type of test. There is a demonstration of a CBT that allows you to experience this type of test before you register. Some technicians find this style of testing provides an advantage, while others find operating the computer a distraction. One significant benefit of CBT is the availability of instant results. You can receive your test results before you leave the test center. CBT also offers increased flexibility in scheduling. The cost for taking CBTs is slightly higher than paper-and-pencil tests and the number of testing sites is limited. The first-time test taker may be more comfortable with the paper-and-pencil tests but technicians now have a choice.

Who writes the questions?

The questions are written by service industry experts in the area being tested. Each area will have its own technical experts. Questions are entirely job related. They are designed to test the skills you need to be a successful technician. Theoretical knowledge is important and necessary to answer the questions, but the ability to apply that knowledge is the basis of ASE test questions.

Each question has its roots in an ASE "item-writing" workshop where service representatives from automobile manufacturers (domestic and import), aftermarket parts and equipment manufacturers, working technicians, and vocational educators meet in a workshop setting to share ideas and translate them into test questions. Each test question written by these experts must survive review by all members of the group.

The questions are written to deal with practical application of soft skills and system knowledge experienced by technicians in their day-to-day work.

All questions are pre-tested and quality-checked on a national sample of technicians. Those questions that meet ASE standards of quality and accuracy are included in the scored sections of the tests; the "rejects" are sent back to the drawing board or discarded altogether.

Each certification test is made up of between 40 and 80 multiple choice questions.

Note: Each test could contain additional questions that are included for statistical research purposes only. Your answers to these questions will not affect your score, but since you do not know which ones they are, you should answer all questions on the test. The five-year Recertification Test will cover the same content areas as those previously listed. However, the number of questions in each content area of the Recertification Test will be reduced by about one-half.

Objective Tests

A test is called an objective test if the same standards and conditions apply to everyone taking the test and there is only one correct answer to each question.

Objective tests primarily measure your ability to recall information. A well-designed objective test can also test your ability to understand, analyze, interpret, and apply your knowledge. Objective tests include true-false, multiple choice, fill in the blank, and matching questions. ASE's tests consist exclusively of four-part multiple choice objective questions.

The following are some strategies that may be applied to your tests.

Before taking an objective test, quickly look over the test to determine the number of questions, but do not try to read through all of the questions. In an ASE test, there are usually between 40 and 80 questions, depending on the subject. Read through each question before marking your answer. Answer the questions in the order they appear on the test. Leave the questions blank that you are not sure of and move on to the next question. You can return to those unanswered questions after you have finished the others. They may be easier to answer at a later time after your mind has had additional time to consider them on a subconscious level. In addition, you might find information in other questions that will help you recall the answers to some of them.

Do not be obsessed by the apparent pattern of responses. For example, do not be influenced by a pattern like **D**, **C**, **B**, **A**, **D**, **C**, **B**, **A** on an ASE test.

There is also a lot of folk wisdom about taking objective tests. For example, there are those who would advise you to avoid response options that use certain words such as *all, none, always, never, must,* and *only,* to name a few. This, they claim, is because nothing in life is exclusive. They would advise you to choose response options that use words that allow for some exception, such as *sometimes, frequently, rarely, often, usually, seldom,* and *normally.* They would also advise you to avoid the first and last option (A and D) because test writers, they feel, are more comfortable if they put the correct answer in the middle (B and C) of the choices. Another recommendation often offered is to select the option that is either shorter or longer than the other three choices because it is more likely to be correct. Some would advise you to never change an answer since your first intuition is usually correct.

Although there may be a grain of truth in this folk wisdom, ASE test writers try to avoid this and so should you. There are just as many **A** answers as there are **B** answers, just as many **D** answers as **C** answers. As a matter of fact, ASE tries to balance the answers at about 25 percent per choice **A**, **B**, **C**, and **D**. There is no intention to use "tricky" words, such as previously outlined. Put no credence in the opposing words "sometimes" and "never," for example.

Multiple choice tests are sometimes challenging because there are often several choices that may seem possible, and it may be difficult to decide on the correct choice. The best strategy, in this case, is to determine the correct answer before looking at the options. If you see the answer you decided on, you should still examine the options to make sure that none seem more correct than yours. If you do not know or are not sure of the answer, read each option very carefully and try to eliminate those options that you know to be wrong. That way, you can often arrive at the correct choice through a process of elimination.

If you have gone through all of the test and you still do not know the answer to some of the questions, then guess. Yes, guess. You then have at least a 25 percent chance of being correct. If you leave the question blank, you have no chance. Your score is based on the number of questions answered correctly.

Preparing for the Exam

The main reason we have included so many sample and practice questions in this guide is, simply, to help you learn what you know and what you don't know. We recommend that you work your way through each question in this book. Before doing this, carefully look through Section 3; it contains a description and explanation of the question types you'll find on an ASE exam.

Once you understand what the questions will look like, move to the sample test. Answer one of the sample questions (Section 5) then read the explanation (Section 7) to the answer for that question. If you don't feel you understand the reasoning for the correct answer, go back and read the overview (Section 4) for the task that is related to that question. If you still don't feel you have a solid understanding of the material, identify a good source of information on the topic, such as a textbook, and do some more studying.

After you have completed all of the sample test items and reviewed your answers, move to the additional questions (Section 6). This time answer the questions as if you were taking an actual test. Do not use any reference or allow any interruptions in order to get a feel for how you will do on an actual test. Once you have answered all of the questions, grade your results using the answer key in Section 7. For every question that you gave a wrong answer to, study the explanations to the answers and/or the overview of the related task areas. Try to determine the root cause for your missing the question. The easiest thing to correct is learning the correct technical content. The hardest things to correct are behaviors that lead you to a wrong conclusion. If you knew the information but still got it wrong there is a behavior problem that will need to be corrected. An example would be reading too quickly and skipping over words that affect your reasoning. If you can identify what you did that caused you to answer the question incorrectly you can eliminate that cause and improve your score. Here are some basic guidelines to follow while preparing for the exam:

- Focus your studies on those areas in which you are weak.
- Be honest with yourself while determining if you understand something.
- Study often but in short periods of time.
- Remove yourself from all distractions while studying.
- Keep in mind the goal of studying is not just to pass the exam, the real goal is to learn!
- Prepare physically by getting a good night's rest before the test and eat meals that provide energy but do not cause discomfort.
- Arrive early to the test site to avoid long waits as test candidates check in and to allow all of the time available for your tests.

During the Test

On paper-and-pencil tests you will be placing your answers on a sheet where you will be required to color in your answer choice. Stray marks or incomplete erasures may be picked up as an answer by the electronic reader, so be sure only your answers end up on the sheet. One of the biggest problems an adult faces in test taking, it seems, is placing the answer in the correct spot on the answer sheet. Make certain that you mark your answer for, say, question 21, in the space on the answer sheet designated for the answer for question 21. A correct response in the wrong line will probably result in two questions being marked wrong, one with two answers (which could include a correct answer but will be scored wrong) and the other with no answer. Remember, the answer sheet on the written test is machine scored and can only "read" what you have colored in.

If you finish answering all of the questions on a test and have remaining time, go back and review the answers to those questions that you were not sure of. You can often catch careless errors by using the remaining time to review your answers. Carefully check your answer sheet for blank answer blocks or missing information.

At practically every test, some technicians will invariably finish ahead of time and turn their papers in long before the final call. Some technicians may be doing recertification tests and others may be taking fewer tests than you. Do not let them distract or intimidate you.

It is not wise to use less than the total amount of time that you are allotted for a test. If there are any doubts, take the time for review. Any product can usually be made better with some additional effort. A test is no exception. It is not necessary to turn in your test paper until you are told to do so.

Testing Time Length

An ASE written test session is four hours. You may attempt from one to a maximum of four tests in one session. It is recommended, however, that no more than a total of 225 questions be attempted at any test session. This will allow for just over one minute for each question.

Visitors are not permitted at any time. If you wish to leave the test room for any reason, you must first ask permission. If you finish your test early and wish to leave, you are permitted to do so only during specified dismissal periods.

You should monitor your progress and set an arbitrary limit to how much time you will need for each question. This should be based on the number of questions you are attempting. It is suggested that you wear a watch because some facilities may not have a clock visible to all areas of the room.

Computer-based tests are allotted a testing time according to the number of questions ranging from one half hour to one and one half hours. Advanced level tests are allowed two hours. This time is by appointment and you should be sure to be on time to ensure that you have all of the time allocated. If you arrive late for a CBT test appointment you will only have the amount of time remaining on your appointment.

Your Test Results!

You can gain a better perspective about tests if you know and understand how they are scored. ASE's tests are scored by American College Testing (ACT), a nonpartial, unbiased organization having no vested interest in ASE or in the automotive industry.

Each question carries the same weight as any other question. For example, if there are 50 questions, each is worth 2 percent of the total score.

The test results can tell you:

- where your knowledge equals or exceeds that needed for competent performance.
- where you might need more preparation.

Your ASE test score report is divided into content areas and will show the number of questions in each content area and how many of your answers were correct. These numbers provide information about your performance in each area of the test. However, because there may be a different number of questions in each content area of the test, a high percentage of correct answers in an area with few questions may not offset a low percentage in an area with many questions.

It should be noted that one does not “fail” an ASE test. The technician who does not pass is simply told “More Preparation Needed.” Though large differences in percentages may indicate problem areas, it is important to consider how many questions were asked in each area. Since each test evaluates all phases of the work involved in a service specialty, you should be prepared in each area. A low score in one area could keep you from passing an entire test.

There is no such thing as average. You cannot determine your overall test score by adding the percentages given for each task area and dividing by the number of areas. It doesn’t work that way because there generally are not the same number of questions in each task area. A task area with 20 questions, for example, counts more toward your total score than a task area with 10 questions.

Your test report should give you a good picture of your results and a better understanding of your strengths and weaknesses for each task area.

If you fail to pass the test, you may take it again at any time it is scheduled to be administered. You are the only one who will receive your test score. Test scores will not be given over the telephone by ASE nor will they be released to anyone without your written permission.

3 Types of Questions on an ASE Exam

ASE certification tests are often thought of as being tricky. They may seem to be tricky if you do not completely understand what is being asked. The following examples will help you recognize certain types of ASE questions and avoid common errors.

Paper-and-pencil tests and computer-based test questions are identical in content and difficulty. Most initial certification tests are made up of 40 to 80 multiple choice questions. Multiple choice questions are an efficient way to test knowledge. To answer them correctly, you must think about each choice as a possibility, and then choose the one that best answers the question. To do this, read each word of the question carefully. Do not assume you know what the question is about until you have finished reading it.

About 10 percent of the questions on an actual ASE exam will use an illustration. These drawings contain the information needed to correctly answer the question. The illustration must be studied carefully before attempting to answer the question. Often, technicians look at the possible answers then try to match up the answers with the drawing. Always do the opposite; match the drawing to the answers. When the illustration is showing an electrical schematic or another system in detail, look over the system and try to figure out how the system works before you look at the question and the possible answers.

Multiple Choice Questions

The most common type of question used on ASE tests is the multiple choice question. This type of question contains three "distracters" (wrong answers) and one "key" (correct answer). When the questions are written effort is made to make the distracters plausible to draw an inexperienced technician to one of them. This type of question gives a clear indication of the technician's knowledge. Using multiple criteria including cross-sections by age, race, and other background information, ASE is able to guarantee that a question does not bias for or against any particular group. A question that shows bias toward any particular group is discarded. If you encounter a question that you are unsure of, reverse engineer it by eliminating the items that it cannot be. For example:

Under load, a diesel engine emits dark black smoke from the exhaust pipe. What should the technician check first?

A. air cleaner
B. fuel pump
C. injector pump
D. turbocharger (A1)

Analysis:

Answer A is correct. Whenever a diesel engine emits black smoke, oxygen starvation caused by air cleaner restriction should be checked first, mainly because it can be checked and quickly eliminated as a cause. Test the intake air inlet restriction using a water manometer. The specifications should always be checked to the OEM values, but typical maximum values will be close to:

15 inches H_2O vacuum naturally aspirated engines
25 inches H_2O vacuum boosted engines

Answer B is wrong. The fuel pump (transfer pump) is very unlikely to cause a black smoke condition.

Answer C is wrong. The injector pump may cause black smoke, but it isn't the first item to check.
Answer D is wrong. The turbocharger can cause black smoke, but it isn't the first item to check.

EXCEPT Questions

Another type of question used on ASE tests has answers that are all correct except one. The correct answer for this type of question is the answer that is wrong. The word "EXCEPT" will always be in capital letters. You must identify which of the choices is the wrong answer. If you read quickly through the question, you may overlook what the question is asking and answer the question with the first correct statement. This will make your answer wrong. An example of this type of question and the analysis is as follows:

All of the following may cause premature clutch disc failure EXCEPT

A. oil contamination of the disc.
B. worn torsion springs.
C. worn U-joints.
D. a worn clutch linkage. (A1)

Analysis:

Answer A is wrong. Oil contamination may cause slippage and disc failure.
Answer B is wrong. Worn torsion springs may cause clutch disc hub damage.
Answer C is correct. Worn U-joints will not cause premature clutch failure. This may cause driveline noise and vibration.
Answer D is wrong. Worn clutch linkage may cause disc failure due to incomplete clutch engagement or disengagement.

Technician A, Technician B Questions

The type of question that is most popularly associated with an ASE test is the "Technician A says . . . Technician B says . . . Who is correct?" type. In this type of question, you must identify the correct statement or statements. To answer this type of question correctly, you must carefully read each technician's statement and judge it on its own merit to determine if the statement is true.

Sometimes this type of question begins with a statement about some analysis or repair procedure. This is often referred to as the stem of the question and provides the setup or background information required to understand the conditions the question is based on. This is followed by two statements about the cause of the concern, proper inspection, identification, or repair choices. You are asked whether the first statement, the second statement, both statements, or neither statement is correct. Analyzing this type of question is a little easier than the other types because there are only two ideas to consider although there are still four choices for an answer.

Technician A, Technician B questions are really double true or false questions. The best way to analyze this kind of question is to consider each technician's statement separately. Ask yourself, is A true or false? Is B true or false? Then select your answer from the four choices. An important point to remember is that an ASE Technician A, Technician B question will never have Technician A and B directly disagreeing with each other. That is why you must evaluate each statement independently.

An example of this type of question and the analysis of it follows.

When discussing the adjustment of a single disc push-type clutch, Technician A says that the clutch adjustment is within the pressure plate. Technician B says that the adjustment is through the linkage only. Who is correct?

A. A only
B. B only
C. Both A and B
D. Neither A nor B (A5)

Analysis:

Answer A is wrong. There is no provision for adjustment in the pressure plates of push-type clutches.
Answer B is correct. Only Technician B is correct. All adjustment is done through the linkage on a push-type clutch.
Answer C is wrong. Only Technician B is correct.
Answer D is wrong. Only Technician B is correct.

Most-Likely Questions

Most-Likely questions are somewhat difficult because only one choice is correct while the other three choices are nearly correct. An example of a Most-Likely-cause question is as follows:

In a twin countershaft transmission, a noise is noticeable in all gear shift positions except for high gear (direct). The Most-Likely cause of this noise is

A. a worn countershaft gear.
B. worn countershaft bearings.
C. worn rear main shaft support bearings.
D. worn front main shaft support bearings. (B2)

Analysis:

Answers A is wrong. A worn countershaft gear would make noise whenever rotated, which is whenever the input shaft is turning.
Answer B is wrong. Worn countershaft bearings would make noise whenever rotated, which is whenever the input shaft is turning.
Answer C is wrong. A worn rear main shaft support bearing would make a noise whenever the main shaft turns, which is in every gear.
Answer D is correct. The front main shaft support bearing will only make noise when there is a speed difference between the main shaft and input shaft, which is in every gear but direct. The truck should be road tested to determine if the driver's complaint of noise is actually in the transmission. Also, technicians should try to locate and eliminate noise by means other than transmission removal or overhaul. If the noise does seem to be in the transmission, try to break it down into classifications. If possible, determine what position the gearshift lever is in when the noise occurs. If the noise is evident in only one gear position, the cause of the noise is generally traceable to the gears in operation. Jumping out of gear is usually caused by excessive end play on gears or synchronizer assemblies. This problem may also be caused by weak or broken detent springs and worn detents on the shifter rails.

LEAST-Likely Questions

Notice that in Most-Likely questions there is no capitalization. This is not so with LEAST-Likely type questions. For this type of question, look for the choice that would be the LEAST-Likely cause of the described situation. Read the entire question carefully before choosing your answer. An example is as follows:

When diagnosing an electronically controlled, automated mechanical transmission, which tool would be LEAST-Likely used?

A. a digital volt, ohmmeter
B. a laptop computer
C. a handheld scan tool
D. a test light (B7)

Analysis:

Answers A is wrong. Digital multimeters are commonly used by technicians for diagnosing electronically controlled transmissions.
Answer B is wrong. Laptop computers are commonly used by technicians for diagnosing electronically controlled transmissions.
Answer C is wrong. A hand-held scan tool is commonly used for diagnosing electronically controlled transmissions.
Answer D is correct. A test light would be the least likely tool used for diagnosing. Test lights should not be used when diagnosing because these are not high impedance tools and may cause damage to sensitive electronic components.

Summary

There are no four-part multiple choice ASE questions having "none of the above" or "all of the above" choices. ASE does not use other types of questions, such as fill-in-the-blank, completion, true-false, word-matching, or essay. ASE does not require you to draw diagrams or sketches. If a formula or chart is required to answer a question, it is provided for you. There are no ASE questions that require you to use a pocket calculator.

Overview of Task List

Heating, Ventilation and Air Conditioning (HVAC) Systems (Test T7)

The following section includes the task areas and task lists for this test and a written overview of the topics covered in the test.

The task list describes the actual work you should be able to do as a technician that you will be tested on by the ASE. This is your key to the test and you should review this section carefully. We have based our sample test and additional questions upon these tasks and the overview section will also support your understanding of the task list. ASE advises that the questions on the test may not equal the number of tasks listed; the task lists tell you what ASE expects you to know how to do and be ready to be tested on.

At the end of each question in the Sample Test and Additional Test Questions Sections, a letter and number will be used as a reference back to this section for additional study. Note the following example: **B1.2.**

B. A/C System and Component Diagnosis, Service, and Repair (16 questions)

1. A/C System. General (6 questions)

Task B1.2 Identify refrigerant type and check for contamination; determine appropriate action.

Example:

1. Which of these characteristics does the R-134a refrigerant possess?
   A. odorless
   B. a faint ether-like odor
   C. a strong rotten egg odor
   D. a cabbage-like odor (B1.2)

Analysis:

Question #1
Answer A is wrong. R-134a does have an ether-like odor.
Answer B is correct. R-134a has a faint ether-like odor. It is possible for a customer to comment that they have an ether-like smell in their vehicle if the evaporator core has a large leak.
Answer C is wrong. R-134a does not have a strong rotten egg odor.
Answer D is wrong. R-134a does not have a cabbage-like odor.

Task List and Overview

A. HVAC Systems Diagnosis, Service, and Repair (6 questions)

Task A1 Verify the complaint, road test the vehicle, review driver/customer interview and past maintenance documents (if available); determine further diagnosis.

A service technician's job is to try to repair the vehicle in the most efficient manner possible. To accomplish this, it is advisable to have a good strategy that is followed for each vehicle worked. The first step of repairing the vehicle is to make sure that there is a real problem. In order to verify that there is a problem, the technician needs to attempt to operate the vehicle in a manner that tests the system in question. Many times this requires the technician to road test the vehicle. After verifying that there really is something to fix, a thorough technician gathers as much information as possible to find out about the service history of the truck and also as many details about when, where, and how the problem started. Having this arsenal of information helps the technician understand the problem better and gives him a much higher chance to repair the truck on the first visit.

Task A2 Verify the need of service or repair of HVAC systems based on unusual operating noises; determine appropriate action.

The service technician must be aware of normal HVAC system operating noises in order to determine whether a system requires service. Normal noises include the sounds of A/C compressor clutch engagement, the blower motor, moving blend air and mode doors, and pressure equalization after the vehicle is shut down. Noises that could indicate the need for service include a growling sound from the water pump or A/C compressor, a whistling noise under the dash, or a grinding noise when control levers are moved.

A loose, dry, or worn A/C compressor belt will cause a squealing noise. This noise will be worse during acceleration. Worn or dry blower motor bearings may cause a squealing noise when the blower is running; this noise may occur when the engine first starts after sitting overnight. A loose or worn clutch hub or loose compressor mounting bolt will also cause a rattling noise from the compressor.

If liquid refrigerant enters the compressor, a thumping, banging noise will result. Heavy knocking compressor noises come from the following: refrigerant system blockage, incorrect pressures, or internal damage. A worn compressor pulley bearing or air clutch bearing will cause a growling noise with the compressor disengaged. If the growling noise only occurs when the system engages the clutch, internal bearings may be at fault.

Task A3 Verify the need of service or repair of HVAC systems based on unusual visual, smell, and touch conditions; determine appropriate action.

The service technician must be aware of abnormal conditions in the HVAC system in order to determine the need for system service. If the driver complains of high or low temperatures inside the cab, this is cause for a system performance test. Abnormal conditions include the smell of antifreeze inside the cab, a fogged windshield, ice buildup on A/C components, and oil or dirt buildup on A/C fittings.

Task A4 Identify system type and components (cycling clutch orifice tube CCOT, expansion valve, type of refrigerant) and conduct performance test(s) on HVAC systems; determine appropriate action.

First, the technician must know if the system operates on R-12 or R-134a refrigerant. Most major R-134a components use a light-blue label to indicate an R-134a system. R-12 systems use Schrader-type service valves, where R-134a systems use metric threads and quick connect/disconnect service valves.

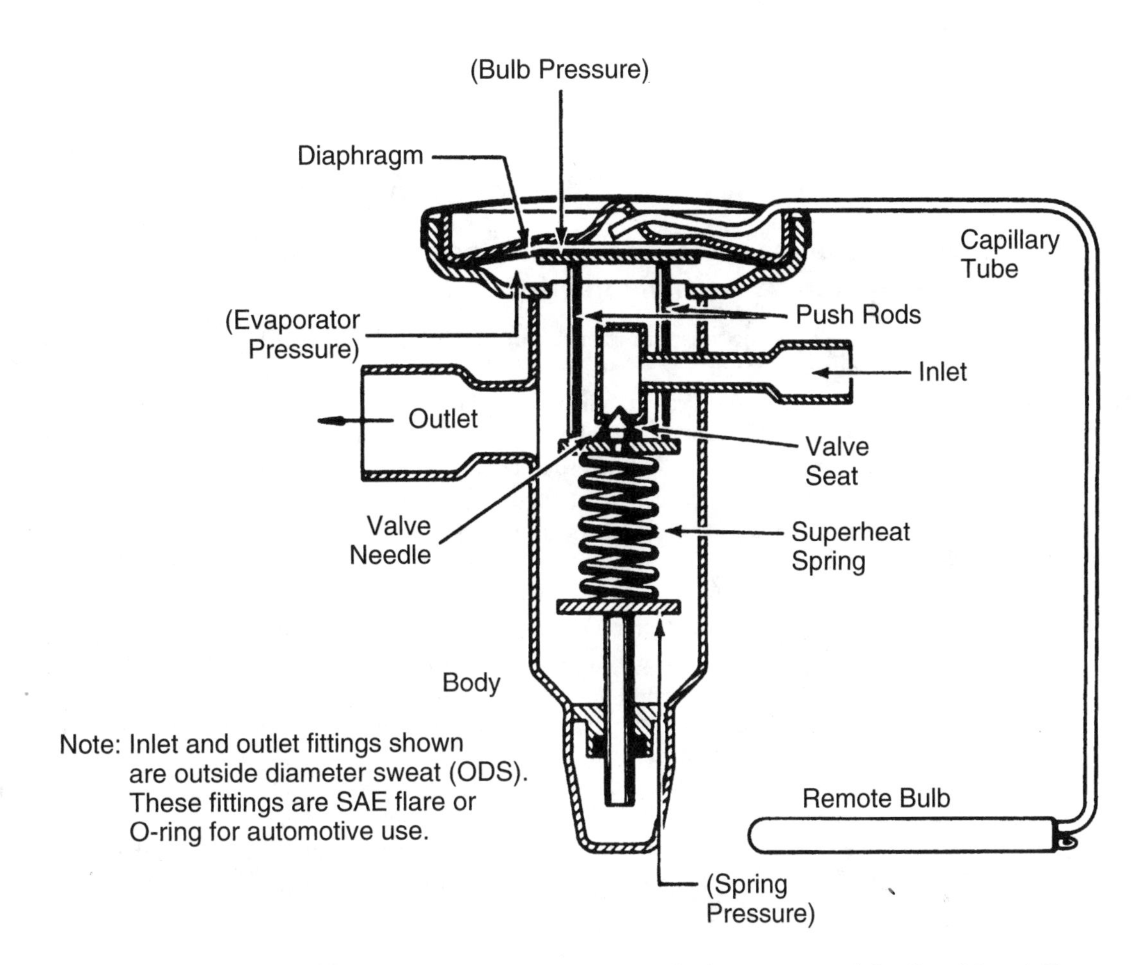

Next, you need to identify the type of expansion device in the evaporator inlet line. Most A/C systems use a TXV as shown in the figure or a fixed orifice tube (FOT) in the evaporator inlet line to control refrigerant flow into the evaporator. You have high pressure liquid on the inlet side of the expansion device and low pressure liquid on the outlet side. Mack trucks use a block-type assembly that contains an equalized TXV.

The truck industry uses three types of A/C systems: mechanical, semiautomatic, or automatic. Mechanical systems use a slide-type lever or rotary switch to control the in-cab temperature manually. In a semiautomatic system, a computer electronically controls outlet air temperature only, but in the automatic systems, most sub-systems are computer controlled.

Manual systems rely on the driver to select the temperature, mode, and blower speed. Semiautomatic temperature control (SATC) systems regulate only the temperature of the output air and rely on the driver to select the desired mode and blower speed. A microprocessor (computer) controls a fully automatic temperature control (ATC) system. These systems use the input from various sensors throughout the vehicle to control the blend doors automatically and to adjust the interior temperature using an appropriate blower speed.

An HVAC performance test should include operation of the system in all modes, and at a variety of temperatures and blower speeds. A small pocket thermometer that is known to be accurate should be used to verify that the output air temperatures match the temperature settings.

B. A/C System and Component Diagnosis, Service, and Repair (16 questions)

1. A/C System. General (6 questions)

Task B1.1 Diagnose the cause of temperature control problems in the A/C system; determine needed repairs.

To check for causes of temperature control problems, begin by checking for compressor clutch engagement when A/C is selected. You should hear the clutch engage and notice a change in engine RPM and sound. If you are unable to tell by this method, watch the compressor clutch while an assistant turns on the A/C. If the clutch fails to engage, check the wiring schematic for the system to identify the power and ground sources and any compressor control devices used. Follow the manufacturer's procedures to pinpoint the source of the failure. If the compressor engages, a quick check to see if the system is operating can be made by comparing the compressor suction and discharge temperatures. The suction side should be cool to the touch and the discharge should be hot. Perform further testing by attaching a manifold gauge set and doing a performance test. Compare to the manufacturer's specifications. After verifying that refrigerant temperatures and pressures are acceptable, check for air temperature control system problems.

The temperature of A/C system output air is generally controlled by one of two ways. (1) In blend air systems, air cooled by the evaporator core is mixed with air warmed by the heater core. Ultimately, the output air temperature control occurs by regulating the amount of air allowed to flow through each core. (2) In other A/C systems, temperature control is achieved by opening and closing the water control valve. When the temperature selector is in the full cold position, the water control valve totally blocks all hot coolant from entering the heater core. When the temperature selector is moved toward the hot setting, the water control valve opens to allow hot coolant into the heater core resulting in warmer air being delivered to the cab. This type of system does not have a blend air door. This type of system typically has the evaporator core and the heater core stacked right next to each other. A/C system outlet air temperature is affected by outside air temperature and humidity, engine coolant temperature, air flow through the condenser and evaporator, and level of refrigerant charge. Outlet air temperature may also be affected by mechanical or electrical failure of system components.

Task B1.2 Identify refrigerant type and check for contamination; determine appropriate action.

The easiest way to identify the type of refrigerant that is used in a given A/C system is to observe the service fittings. Society of Automotive Engineers (SAE) standard J639 defines the size and type of service fittings for R-12 and R-134a A/C systems. R-12 service fittings have external threads. R-134a systems use quick-disconnect fittings. You cannot vent either R-12 or R-134a to the atmosphere. A refrigerant identifier machine may be connected to all types of A/C systems so as to determine refrigerant type and level of contamination.

Note: Recovery machines can be damaged by recovery of chemicals not compatible with their seals and valves.

Task B1.3 Diagnose A/C system problems indicated by pressure gauge readings and sight glass/moisture indicator conditions (where applicable); determine needed service or repairs.

In a normally operating A/C system, the low-side pressure varies between 20 and 45 psi, and the high-side pressure varies between 120 and 300 psi, depending on the ambient temperature and humidity levels. The following table summarizes abnormal A/C system pressures and common causes.

A/C Pressure Diagnosis

LOW-SIDE PRESSURE	HIGH-SIDE PRESSURE	POSSIBLE CAUSES
LOW	LOW	Low refrigerant charge
LOW	LOW	Obstruction in the suction line
LOW	LOW	Clogged orifice tube
LOW	LOW	TXV valve stuck closed*
LOW	LOW	Restricted line from the condenser to the evaporator*
LOW	HIGH	Restricted evaporator air flow
HIGH	LOW	Internal compressor damage
HIGH	HIGH	Refrigerant overcharge
HIGH	HIGH	Restricted condenser air flow
HIGH	HIGH	High engine coolant temperature
HIGH	HIGH	TXV valve stuck open
HIGH	HIGH	Air or moisture in the refrigerant

*Stuck closed TXV valves or a restricted line from the condenser to the evaporator will cause frosting at the point of restriction.

In some A/C systems, a sight glass allows the service technician to make a quick assessment of the system condition. With the A/C compressor clutch engaged, a properly charged system will occasionally show traces of bubbles. A sight glass that appears foamy indicates that the refrigerant charge is low. When the sight glass contains bubbles and/or foam, the refrigerant charge is low and air has entered the system. Oil streaks appearing in the sight glass indicate that compressor oil is circulating through the system. A cloudy sight glass indicates that the desiccant pack in the receiver/drier has broken down.

On R-12 systems, sight glass indications are only valid when the ambient (surrounding area) temperature is above 71°F (21°C). If the temperature is below 70°F, it is normal for bubbles to appear in the sight glass. A clear sight glass may indicate the proper refrigerant charge. A clear sight glass may also indicate an excessive refrigerant charge or no refrigerant charge. Many R-134a systems do not have a sight glass. Most manufacturers agree that the following conditions must be present when performance testing an R-134a system:

- Temperature control set in the lowest position
- High blower speed
- Engine speed set at 1500 rpm or peak-governed rpm
- Recirculation air set
- Monitor low-side and high-side pressure
- Monitor the temperature of the air from the center duct

Most manufacturers have eliminated the sight glass as a diagnostic means, due to its unreliability when used in R-134a visual interpretations. Sometimes it is left in place in the system, as the manufacturer did not wish to redesign that section, but is covered by tape. Leave the tape in place and disregard the sight glass.

Task B1.4 Diagnose A/C system problems indicated by unusual operating noise, visual, smell, and touch procedures; determine needed repairs.

To diagnose A/C systems efficiently, the service technician must be observant. Restricted hoses cause a frosting or sudden temperature change at a specific point along the hose. Frost on the receiver/dryer usually indicates an internal restriction in that component. Because the receiver/dryer is located in the high pressure liquid line between the condenser and evaporator, it should feel warm. Frost formation on the evaporator outlet indicates a flooded evaporator, caused by an excessive refrigerant charge, or a stuck open TXV valve. Unusual noises can often guide the technician to a faulty component. These problems may also cause frost on the compressor suction line.

On systems using an accumulator, it should feel cold because of its close connection to the evaporator. Both the evaporator inlet and outlet should feel cold, when operating normally. On orifice tube systems, the evaporator inlet should be slightly warmer to the touch than the outlet. If the evaporator outlet is warm, the refrigerant charge may be low. High-side refrigerant components should feel hot or warm, and low-side components should feel cool or cold. A plugged drain in the evaporator case causes a strong rotten egg smell in the cab.

A thorough visual inspection is always a good first step in diagnosing A/C systems.

Task B1.5 Perform A/C system leak test; determine needed repairs.

Refrigerant systems use the following two methods to detect leaks: dye check or electronic leak detector. To check an A/C system for leaks, the technician must first ensure that the system contains enough refrigerant to allow compressor clutch engagement. If the system is empty, you should first pressure test the system using nitrogen, then when you are sure the system will hold pressure, install a partial refrigerant charge. When the partial charge is installed, a small amount of dye should also be added. After running with the dye installed for fifteen minutes, the dye will appear at the leak area. Ultraviolet dye is available for installation into the refrigerant and is visible using a black-light detector. Electronic detectors provide an audible beeping sound when the probe is placed near the leak source. Since R-12 and R-134a are different chemically, a specific electronic detector or one that does both systems is used. When checking for leaks, place the leak detector probe directly below each fitting and each component, directly below the evaporator drain, and at the center panel duct. Check the entire system to rule out multiple leaks.

When using an electronic leak detector, calibrate the detector before each use. Other common means of leak detection are a soap and water solution applied to external areas of the system. If leaks are present under pressure, then bubbles will appear. Another method is fluorescent soap; as with the dye, it will change color when it comes into contact with the refrigerant.

Under no circumstances is a flame-type tester to be used. When R-12 comes into contact with a flame, it becomes phosgene gas. Also, all refrigerants will combust at the appropriate temperature.

Task B1.6 Evacuate A/C system using appropriate equipment.

Before you can evacuate a system, you must first recover any remaining refrigerant; it is important to follow the manufacturer's instructions for the specific recovery station used. Never vent refrigerant to the atmosphere; it is illegal and environmentally irresponsible.

When the system is completely empty, connect the manifold gauge center hose to a vacuum pump. Operate this pump for 30 minutes with the service valves open and the low-side gauge valve open. After five minutes of operation, the low-side gauge should indicate 20 in. Hg (67.6 kPa), and the high-side should read below zero, unless it is restricted by a stop pin. If the high-side gauge does not drop below zero, this indicates refrigerant blockage. When the technician locates a blockage, this must be repaired before proceeding with the evacuation process. After fifteen minutes of evacuation, the low-side should indicate 24 to 26 in. Hg (81–88 kPa), if there are no leaks. If less than this value, close the low-side gauge valve and observe the gauge. If the low-side gauge needle rises slowly, this indicates a refrigerant leak. Fix the leak and proceed to evacuate the system to at least 27 in. Hg. Most manufacturers recommend that the vacuum pump run for at least 30 minutes to ensure that all moisture is removed from the system.

Task B1.7 Internally clean contaminated A/C system components and hoses.

If desiccant pack deterioration or catastrophic compressor failure occurs, clean all refrigeration system components internally. Internal cleaning of A/C system components is best accomplished by flushing with A/C flush solvent. Before flushing, the compressor and all restricting components and filters must be removed from the system. It is important to regulate the pressure from the nitrogen supply tank to normal system pressure for each component.

Rather than system flushing, many truck manufacturers recommend using an in-line filter between the condenser and the evaporator to remove debris. These in-line filters come with or without a fixed orifice tube (FOT). If you use the filter that contains an FOT, you must remove the other FOT from the system.

Task B1.8 Charge A/C system with correct type and quantity of refrigerant.

The technician must complete the recovery and evacuation procedure before charging a refrigerant system. Modern recovery and charging stations do not require the A/C system to operate during system charging. It is always best to read all of the manufacturer's instructions for the specific charging station used. The original equipment manufacturer (OEM) may recommend high-side (liquid) or low-side (vapor) charging procedures. You must close both the high-side and low-side manifold gauge valves.

Connect the center hose to the proper refrigerant container and open the container valve. With the engine not running, using the high-side (liquid) charging process, open the high-side gauge valve and observe the low-side gauge, then close the high-side gauge. If the low-side gauge does not move from a vacuum to a pressure, this means the refrigerant system is restricted. With no restriction present, open the high-side gauge valve to proceed with the high-side (liquid) charging procedure. Charging is complete when the correct weight of refrigerant has entered the system. Turn the compressor over by hand to make sure that no liquid refrigerant is in the compressor. Now start the engine and run an A/C performance test. If the system is charged with the engine running and the A/C turned on, make sure to charge only through the low side.

Task B1.9 Identify lubricant type needed for system application.

All R-12 systems use mineral-based refrigeration oil to lubricate the compressor and prevent internal corrosion of components. Mineral-based oil is not compatible with R-134a systems. R-134a systems use polyalkylene glycol (PAG) oil. The PAG lubricant is a synthetic oil and is not compatible with R-12 systems. Many retrofit kits come with ester oil included with them. If the technician is unable to extract all of the mineral oil during the retrofit process, then the ester oil can be used instead of the PAG oil.

2. Compressor and Clutch (5 questions)

Task B2.1 Diagnose A/C system problems that cause protection devices (pressure, thermal, and electronic) to interrupt system operation; determine needed repairs.

A variety of A/C system protection devices can be used in mobile A/C systems. The low pressure cut-out switch will interrupt compressor operation if system pressure drops to the point that a loss of refrigerant charge to the compressor occurs. The high pressure cut-out switch interrupts compressor operation in case of extremely high system pressure. The binary switch combines the function of the low- and high pressure cut-out switches. Some systems have a high pressure relief valve mounted in the receiver/dryer. This valve opens and relieves system pressure if the pressure exceeds 450 to 550 psi (3100 to 3792 kPa). Condenser air flow restrictions cause these extremely high pressures. In gasoline and diesel engine electronic fuel management systems, the computer operates a relay that supplies voltage to the compressor clutch. All input signals go to the engine computer. In some applications, this includes a refrigerant pressure signal. If this input signals an abnormal low- or high pressure condition, the engine computer will not engage the compressor.

Cycling clutch orifice tube (CCOT) systems use a pressure cycling switch to cycle the compressor off and on, in relation to low-side pressure. This switch is mounted in the accumulator between the evaporator and the compressor. This switch closes and turns on the compressor when the refrigerant pressure is above 46 psi (315 kPa). The dash A/C switch supplies the power to the cycling switch. The pressure switch opens when the system pressure decreases to 25 psi (175 kPa). This cycling action maintains the evaporator temperature at 33°F (1°C).

Some refrigerant systems use a thermostatic clutch cycling switch that cycles the compressor on and off in relation to evaporator outlet temperature.

Task B2.2 Inspect, test, and replace A/C system pressure, thermal, and electronic protection devices.

A/C system pressure protection switches are normally closed and may be tested by checking for continuity using a DVOM or a circuit tester while the system is at normal pressure and temperature conditions. The high-pressure switch opens if air-conditioning pressure exceeds about 425–435 psig to prevent damage and closes when pressure drops to below 200 psig. The normally closed low pressure switch, which is usually located on the accumulator, opens when the low side pressure drops below 20–25 psig. Some systems combine several functions into either a binary or trinary switch. Some systems have a thermal fuse, which blows when the compressor begins overheating. Compressor head temperature switches, when used, are mounted so that they contact the compressor case and when they sense that the temperature is exceeding a certain threshold they will open, turning the compressor clutch off. There is usually a diode, which is placed across the compressor clutch coil to reduce voltage spikes, which are caused when the clutch is cycled. This is to protect other voltage sensitive devices. If it is faulty, it will not prevent the clutch from engaging, but will fail to protect other voltage-sensitive parts of the electrical system.

Task B2.3 Inspect and replace A/C compressor drive belts, pulleys, idlers, and tensioners; adjust drive belts and check alignment.

When inspecting an A/C system, it is important that the technician not overlook the A/C compressor drive belts and pulleys. Drive belt edge wear indicates a misaligned or bent pulley. If the belt is loose or bottomed out in the pulley, the belt may slip and cause inadequate cooling. Cracked, glazed, or frayed belts must be replaced. Use a standard drive belt tension gauge to check and adjust belt tension. An improperly adjusted drive belt will wear or fail prematurely. A drive belt adjusted too loosely may slip and cause belt squealing, especially on acceleration with the A/C on and clutch engaged. A drive belt adjusted too tightly may cause internal engine wear or damage to other belt-driven components. Cracked or bent pulleys must be replaced, not repaired.

Task B2.4 Inspect, test, service, and replace A/C compressor clutch components or assembly.

The A/C compressor clutch assembly allows the A/C compressor to engage and disengage to modulate system pressures. The components of the compressor clutch assembly are:

- the driven plate, which is keyed to the compressor driveshaft.
- the drive plate, which is integral to the drive pulley.
- the clutch bearing, which operates when the clutch is disengaged.
- the clutch coil, which creates the magnetic field that engages the clutch.

Defective compressor clutch bearings will make a growling noise with the engine running and the clutch disengaged. Test the compressor clutch by applying power and ground to the appropriate terminals and watching for clutch engagement. You must inspect the pulley and armature plate frictional surfaces for wear and oil contamination. Check the hub bearing for roughness, grease leakage, and looseness. A driven plate that drags on the drive plate or slips briefly on engagement indicates an improper clutch air gap. Adjust this air gap using shims; remove shims to decrease clearance and increase shims to increase clearance. Check the gap on any clutch service. A technician can perform a voltage drop test while the system is in operation. This is the most practical test method as components are in actual running mode.

Task B2.5 Inspect and correct A/C compressor lubricant level.

Check the A/C compressor lubricant level and adjust it anytime there is evidence of lubricant loss from the system. An excessive amount of oil in a refrigerant system reduces system-cooling efficiency. To check the compressor lubricant level, remove the compressor from the vehicle, drain all refrigeration oil, and refill it to manufacturer's specifications.

R-12 systems require a mineral oil with a YN-9 designation, and R-134a systems with a reciprocating compressor must have a synthetic (PAG) oil designation. Rotary compressors use a different type of synthetic oil. If the oils used become mixed, compressor damage will result.

Task B2.6 **Inspect, test, and replace A/C compressor.**

You can diagnose A/C compressor internal damage using a standard A/C gauge set. Low high-side pressure and high low-side pressure on the manifold gauge set may indicate a defective compressor. A faulty compressor bearing will make a growling noise with the engine running and the compressor clutch engaged. Oil dripping from the front of the compressor indicates a faulty front seal. A rattling noise may be caused by loose compressor mounts. A defective pulley bearing will also cause a growling noise with the clutch disengaged. When replacing A/C seals or O-rings, prelubricate them with the mineral-based refrigeration oil.

Task B2.7 **Inspect, repair, or replace A/C compressor mountings and hardware.**

Damaged A/C compressor mounts or mounting plates can cause drive-belt misalignment, improper drive-belt tension, and compressor vibration. Welding can repair cracked mounts and mounting plates; however, care must be taken to align all parts properly, as serpentine belts require alignment of components to within one degree.

3. Evaporator, Condenser, and Related Components (5 questions)

Task B3.1 **Correct system lubricant level when replacing the evaporator, condenser, receiver/drier or accumulator/drier, and hoses.**

Most A/C system components contain refrigeration oil. When replacing the evaporator, condenser, accumulator, receiver/drier, or A/C hoses, the new component should be drained of oil, and fresh refrigeration oil should be added to manufacturer's specifications.

Task B3.2 **Inspect, repair, or replace A/C system hoses, lines, filters, fittings, and seals.**

You check A/C system hoses for damage and leaks during the course of any A/C system maintenance or inspection. Hoses should be replaced if they are cracked, kinked, abraded, or if the fittings show any signs of abuse. Disassemble leaking fittings and replace the O-rings. New O-rings should be lubricated with the appropriate refrigeration oil. Some manufacturers recommend that mineral oil be used to lubricate the O-rings. Using PAG oil on o-rings will lead to early failure due to the PAG attracting moisture to the joint. It is important to use the nitril o-rings on R-134a systems.

A/C system filters and screens are used to prevent particulate (from corrosion, compressor failure, or desiccant breakdown) from circulating through the A/C system and must be replaced if they are clogged, restricted, or damaged. Some systems have a filter in the line between the condenser and the evaporator and some of these filters contain an orifice tube. You must install this type of filter in the proper direction. Special tools are required to service the spring lock couplings that are used on some hose connections.

Task B3.3 **Inspect A/C condenser for proper air flow.**

The A/C condenser should be checked for proper air flow at regular intervals. During a normal A/C system inspection, any bent condenser fins should be straightened and any debris should be cleaned from the condenser. Debris in the condenser air passages causes excessive high-side and low-side pressures and reduced cooling. This problem may also cause the high pressure relief valve to discharge refrigerant. Additionally, you should check the radiator shutter system for proper operation.

Task B3.4 **Inspect, test, and replace A/C system condenser and mountings.**

If any refrigerant tubes are kinked, cracked, or leaking, the A/C condenser must be replaced. Frost on any of the condenser tubing indicates a refrigerant passage restriction. This condition results in excessive high-side and low-side pressures and inadequate cooling. If you replace the condenser, drain the new component and install fresh refrigeration oil to manufacturer's specifications (typically

1 ounce). Condenser mounts and insulators should be checked for proper alignment and deformation which could cause abrasion and fatigue damage.

Task B3.5 Inspect and replace receiver/drier or accumulator/drier.

Most mobile A/C systems use a receiver/drier or an accumulator/drier to ensure an adequate supply of clean, dry refrigerant to the system. If the receiver/dryer inlet and outlet pipes have a significant temperature difference, the receiver/dryer is restricted. Frost forming on the receiver/dryer indicates an internal restriction. Bubbles and foam in the sight glass indicate rust and moisture contamination. Both of these devices contain a bag of desiccant designed to absorb and hold traces of moisture from the refrigerant. The accumulator is located at the outlet of the evaporator. The receiver/dryer is located just upstream of the system expansion device. The accumulator/drier or receiver/drier must be replaced if the A/C system has remained open to the atmosphere for an extended period of time. Other reasons for replacing the drying device are evidence of moisture or corrosion in the system, or if catastrophic compressor failure has occurred. When the accumulator/drier or receiver/drier is replaced, fresh refrigerant oil must be added to manufacturer's specifications (typically 1 ounce).

Task B3.6 Inspect, test, and replace cab/sleeper refrigerant solenoid, expansion valve(s), thermostatic switch (thermistor); and check placement of thermal bulb (capillary tube).

One type of A/C system expansion device is the thermal expansion valve. The thermal expansion valve senses evaporator temperature using a capillary tube connected to a thermal bulb. As the fluid inside the thermal bulb expands, the orifice in the expansion valve opens to increase refrigerant flow through the evaporator. If the evaporator core temperature drops to near the freezing point, the fluid in the thermal bulb contracts and the expansion valve closes to restrict refrigerant flow. Different designs place this thermal bulb either embedded in the evaporator fins or affixed to the evaporator outlet with insulating tape.

In some systems, the expansion valve is housed in a combination valve or an "H" valve. In these systems, internal mechanisms monitor evaporator inlet and outlet temperatures and pressures and adjust the valve opening accordingly. A technician can diagnose a faulty expansion valve using a set of A/C pressure gauges and consulting a diagnostic chart.

Some A/C systems use a thermostatic switch to monitor evaporator temperature. If the temperature gets too cold, this switch opens to interrupt power to the A/C compressor clutch. These switches are used as de-icing devices in the A/C system.

Task B3.7 Inspect and replace orifice tube.

Cycling clutch-type A/C systems often use a fixed orifice tube as an expansion device. The orifice tube is located at the evaporator inlet and contains a fine screen to prevent the circulation of particulate through the evaporator core and back to the compressor. The orifice tube should be replaced if the screen is restricted or corroded, or in case of desiccant bag breakdown or catastrophic compressor failure.

A restricted orifice tube may cause lower than specified low-side pressure, frosting of the orifice tube, and inadequate cooling from the evaporator. If these conditions appear, place a shop towel soaked in hot water around the orifice tube. If the low-side pressure increases, there is moisture freezing in the orifice tube. To rectify this condition, you must recover, evacuate, and recharge the system. If the hot shop towel did not increase the low-side pressure, clean or replace the orifice tube.

Task B3.8 Inspect, test, and replace cab/sleeper evaporator core.

The A/C evaporator core is located (along with the heater core and air flow control doors) in the evaporator case. One detects a leaking evaporator core most easily by measurement at the evaporator case drain, but it may also be detected at the panel and defroster vents. Also, you can remove the blower resistor assembly and go through that cavity. If the evaporator core has a large leak, an oily film appears on the inside of the windshield and the cab temperature becomes warmer than specified. If you replace an evaporator core, add fresh refrigeration oil to manufacturer's specifications (typically 3 ounces).

Task B3.9 Inspect, clean, and repair evaporator housing and water drain; inspect and service/replace evaporator air filter.

The evaporator case or housing contains the evaporator core, the heater core, the evaporator core drain, and the blend air and mode control doors. A clogged evaporator core drain will cause windshield fogging or a noticeable mist from the panel vents. A clogged drain can normally be opened up with a slender piece of wire or with low pressure shop air. The evaporator drain should be checked during routine maintenance inspections. A cracked evaporator case can cause a whistling noise during high blower operation. Minor cracks can be repaired using epoxy-type adhesives. A mildew smell that is noticeable during A/C system operation can be rectified by removing the evaporator case and washing it with a vinegar and water solution or a commercially available cleaner. Some manufacturers make an aerosol fungicide that can be sprayed onto the evaporator core surface. Access to the evaporator core can sometimes be gained by removing the blower resistor.

Task B3.10 Identify, inspect, and replace A/C system service ports/valves (gauge connections).

Two types of A/C system service ports/valves are used in mobile A/C systems. Older systems use a three-position stem-type valve. Using the figure, B shows the normal operation, or back-seated position, where the valve stem is rotated counterclockwise to seat the rear valve face and seal off the service gauge port. C shows the mid-position used during A/C system diagnosis and service. A shows the front-seated position (valve stem rotated clockwise to seat the front valve face) that isolates the compressor from the A/C system. This position allows a technician to service the compressor without discharging the entire system. The system must never be operated with either service port/valve in the front-seated position.

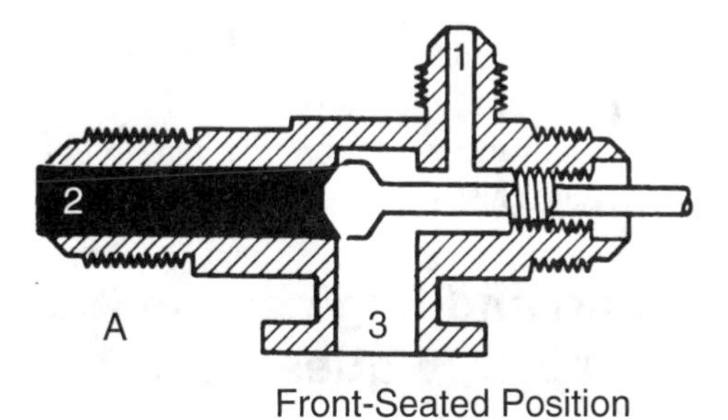

Front-Seated Position

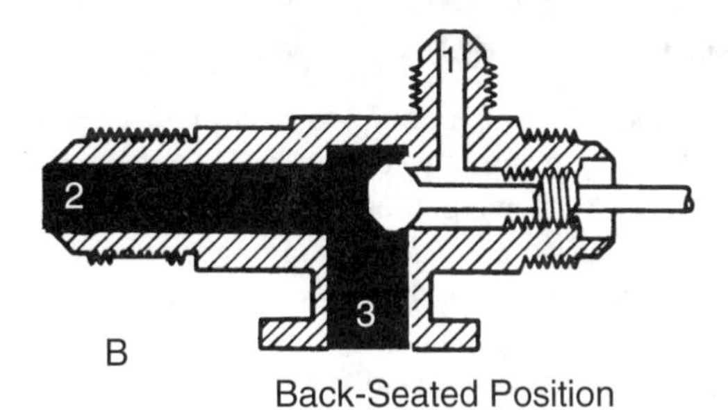

Back-Seated Position

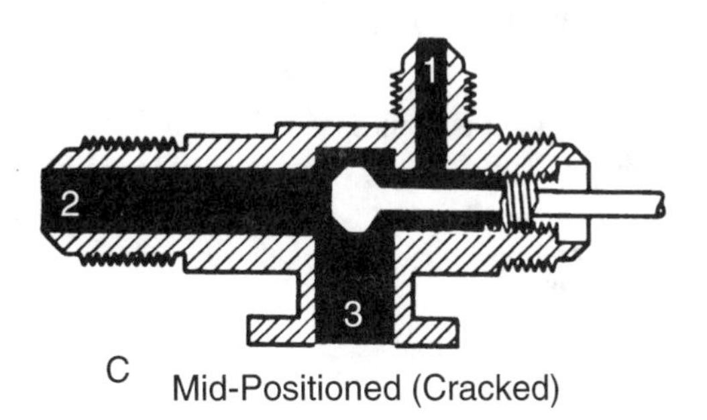

Mid-Positioned (Cracked)

Most R-12 A/C systems have service ports like figure A, which feature external threads and replaceable Schrader valves like the ones used in tire valve stems. R-134a service ports on the other hand, have internal threads and are a quick-disconnect design like figure B. The low-side port is 13mm in diameter and the high-side port is 16mm in diameter.

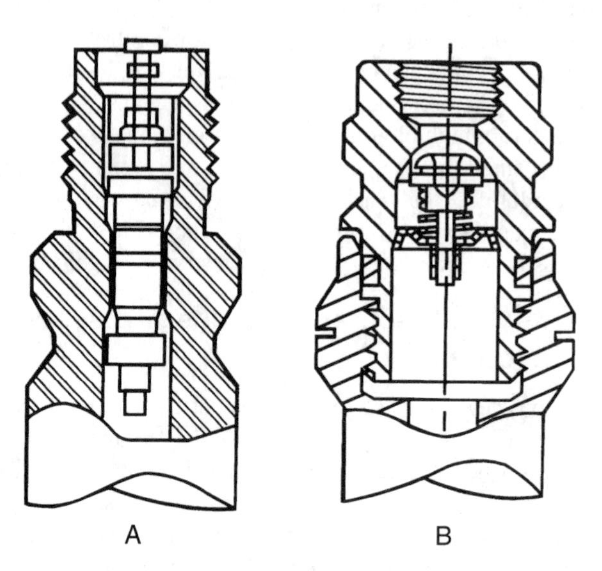

Task B3.11 Diagnose system failures resulting in refrigerant loss from the A/C system high pressure relief device.

All mobile A/C systems are equipped with a high pressure relief device. In most systems, this device is a self-resetting relief valve, which is threaded into the high side of the compressor. The high pressure relief valve will vent refrigerant from the system in the event that high-side pressure exceeds safe levels. This is often an overlooked area when total refrigerant loss has occurred.

C. Heating and Engine Cooling Systems Diagnosis, Service, and Repair (6 questions)

Task C1 Diagnose the cause of outlet air temperature control problems in the HVAC system; determine needed repairs.

Heating system design achieves heater temperature control by one of two methods: blend air modulation or coolant flow control. In a blend air system, coolant flows through the heater core at a constant rate. Air flowing through the HVAC duct is channeled by the blend-air door to flow through or around the heater core, depending on the temperature lever setting. In a coolant flow control system, temperature control is achieved by using a coolant control valve (hot-water valve) to limit the amount of coolant allowed to flow through the heater core, depending on the temperature lever setting. The coolant control valve may be vacuum or cable operated.

Task C2 Diagnose window fogging problems; determine needed repairs.

Windshield fogging may be caused by a leaking heater core or by a clogged evaporator case drain. If windshield fogging is accompanied by the smell of antifreeze, the heater core is at fault. If you smell a pungent odor, then the evaporator case drain may be clogged.

Task C3 Perform engine cooling system tests for leaks, protection level, contamination, coolant level, temperature, coolant type, and conditioner concentration; determine needed repairs.

To test the engine cooling system for leaks, remove the radiator pressure cap following the manufacturer's instructions. Replace the cap with a standard cooling system pressure tester and pressurize the system to operating pressure. Perform a careful visual inspection for leaks. Often you can confirm cooling system contamination with a visual inspection. Rust in the system will turn the coolant an opaque reddish-brown color. If engine oil or transmission fluid has entered the system, the coolant will contain thick deposits resembling a milk shake. Coolant freeze protection level is most easily determined using a cooling system hydrometer or refractometer, which is preferred by some engine manufacturers.

To check the concentration of the cooling system conditioner, you will use test strips that match the type of conditioner being used and then adding conditioner as needed. Be careful not to over-condition the coolant because it can lead to conditioner dropout and system damage. Cooling system temperatures are easily checked using a gauge known to be accurate or a temperature probe.

Task C4 Inspect and replace engine cooling and heating system hoses, lines, fittings, and clamps.

Cooling system and heater hoses must be replaced if they are cracked or brittle, or if they show signs of bulging or abrasion. Hoses begin to deteriorate from the inside first, so examine the inside of any suspected hoses to determine their true condition. Often a cooling system leaks only when cold, which is known as a cold-water leak. The use of constant torque clamps will minimize this. Hose clamps should be replaced if they are deformed or cracked, or if they cannot be operated smoothly.

Task C5 Inspect and test radiator, pressure cap, and coolant recovery system (surge tank); determine needed repairs.

The technician should inspect the radiator at every scheduled preventive maintenance inspection. Check the radiator for bent fins, kinked or cracked tubes, and leaks. With the engine at normal operating temperature, the temperature of the radiator core should be uniform. Cool spots indicate clogged tubes in the core. You test the pressure cap for pressure using a cooling system pressure tester. Make sure the vacuum valve in the cap is working properly. If the pressure cap seal, sealing gasket, or seat is damaged, the engine will overheat and coolant is lost in the recovery system. Replace the cap if it does not hold the pressure specified by the manufacturer or if there is any sign of physical damage. The coolant surge tank should be checked for leaks and for sediment buildup, and should be repaired or cleaned as necessary.

Task C6 Inspect water pump and drive system; determine needed repairs.

The water pump is driven by means of belt drive on light-duty engines and gear driven on heavy-duty applications. The most apparent sign of water pump failure is fluid loss from the weep hole area. A visual inspection of bent, misaligned, or worn pulleys must be performed and corrections made. Gear drives should be checked for clearance (lash) when reassembling.

Task C7 Inspect and test thermostats, by-passes, housings, and seals; determine needed repairs.

The only accurate way to test a thermostat is to remove it from the vehicle, place it in a container of water with a thermometer, and heat the container until the thermostat opens. The water temperature at the point when the thermostat opens should equal the manufacturer's specification for thermostat opening temperature. Other indications that the thermostat has opened include visible coolant flow in the upper radiator tank, a hot upper radiator hose, and an engine temperature gauge that indicates normal operating temperature.

Task C8 Flush and refill cooling system; bleed air from system.

The cooling system should be flushed if there is any sign of rust or contamination in the coolant. After flushing, the entire system should be drained and fresh coolant should be added. The technician should consult the service manual to verify cooling system capacity and bleeding procedure.

Task C9 Inspect, test, and repair or replace fan, fan hub, fan clutch, fan controls, fan thermostat, and fan shroud.

The technician should inspect the engine-cooling fan for loose, cracked, or otherwise damaged blades. Further, inspect the fan hub for cracks. The cooling fan should be replaced, not repaired, if any damage is found. The cooling-fan clutch may be a viscous-type clutch or may be operated by a thermostatic spring. Some heavy-duty trucks use a computer-controlled clutch operated by either the chassis air system or engine oil hydraulic controls.

Since modern trucks feature fans that are controlled by a computer, an examination of fans controls must include verifying the operation of the computer input sensors such as the coolant temperature sensor and the oil temperature sensor. Also, examine any computer-controlled relays that are needed for fan operation. Use the OEM recommended electronic service tools (EST) to verify the operation of ECM controlled circuits. Many trucks have a toggle switch on the dash, which allows the driver to manually turn the fan on, and some systems incorporate the fan into their engine braking strategy for increased performance. Other systems use fan thermostats having a temperature-sensing bulb immersed in the coolant, which either open or close depending upon temperature and system design. They may use either air pressure or oil pressure to actuate the fan clutch.

The fan shroud should be inspected for cracks, and replaced as necessary. If fan blade and shroud damage is found, the technician should verify that the engine mounts are in good condition before replacing the shroud and blade.

Task C10 Inspect, test, and replace heating system coolant control valve(s) and manual shut-off valves.

The coolant control valve (hot-water valve) controls the flow of coolant through the heater core. The coolant control valve may be vacuum operated or cable operated. One can verify proper valve operation by manually opening and closing the valve and observing the temperature change in the downstream heater hose.

Task C11 Inspect, flush, and replace heater core.

In a poorly maintained cooling system, sediment may build up in the heater core causing poor heater performance. Flushing the heater core will restore heater efficiency and may reveal small heater core leaks. The heater core can be pressure tested independently from the rest of the cooling system; however, the core should never be pressurized in excess of normal operating pressure. When replacing a heater core, it is important to reinstall all foam mounting insulators to minimize the risk of vibration damage and to ensure a good seal around the heater core.

D. Operating Systems and Related Controls Diagnosis and Repair (8 questions)

1. Electrical (5 questions)

Task D1.1 Diagnose the cause of failures in HVAC electrical control systems; determine needed repairs.

Diagnosing HVAC electrical control system problems is no different than diagnosing other electrical systems. The technician should follow a systematic approach to troubleshooting, including verifying the concern and performing a thorough visual inspection. The technician should use all available resources, including electrical schematic diagrams and service manual diagnostic routines, to locate and repair the cause of the concern.

Task D1.2 Inspect, test, repair, and replace A/C heater blower motors, resistors, switches, relays, modules, wiring, and protection devices.

A 30-amp fuse or circuit breaker generally protects the HVAC blower circuit. Many blower systems are powered by one or more relays. The operator selects the desired blower motor speed by using the blower switch. The blower resistor block contains several resistors in series, and is used to step-down the voltage to the blower motor, thereby providing multiple blower speeds. The resistor block usually contains a thermal fuse to prevent blower motor damage in case of a high current draw.

Task D1.3 Inspect, test, repair, and replace A/C compressor clutch relays, modules, wiring, sensors, switches, diodes, and protection devices.

The A/C compressor clutch coil is usually energized by an electronically controlled relay. The compressor clutch relay may be controlled by the chassis control unit or by the engine control unit. Many blower motor circuits connect the motor switch and resistor branches to the ground side of the motor. When the system is on, voltage is applied to one brush in the motor. The other brush is connected to the resistor assembly. When you set the blower to high, the motor brush grounds directly through the motor switch contacts at the high-speed relay. If you select one of the lower speeds, the brush grounds through that specific blower resistor.

Task D1.4 Inspect and test A/C-related electronic engine control systems; determine needed repairs.

A/C compressor clutch operation may be dependent upon signals from various engine sensors. Defective engine-related components that affect A/C system operation can usually be diagnosed using a hand-held scan tool or a laptop PC interface. The ECM or chassis management module will disable the A/C compressor clutch if the engine coolant temperature is too high or if the outside air temperature is too low. On vehicles with an electronically controlled transmission, the compressor clutch can be disengaged briefly during shifts. The radiator shutter system is disabled (shutters are fully open) during A/C system operation.

Task D1.5 Inspect, test, repair, and replace engine cooling/condenser fan motors, relays, modules, switches, sensors, wiring, and protection devices.

Some trucks are equipped with electric engine cooling fans and electric condenser fans. Electric fans are usually controlled by an electronic relay which is in turn controlled by the ECM or chassis management module. The ECM grounds the low- and high-speed fan relays in response to engine coolant temperature and compressor head temperature.

When the engine coolant temperature reaches a predetermined temperature set by the manufacturer, the ECM grounds the low speed fan relay. If the coolant temperature continues to rise and exceeds another predetermined set point, the high-speed relay is grounded. The operating ranges are application specific, so check the appropriate service manual and make sure that replacement parts have the correct setting. In general, operating ranges are between 198°F and 234°F. Compressor head temperature switches, when used are mounted so that they contact the compressor case, and when they sense that the temperature is exceeding a certain threshold, will open, turning the compressor clutch off.

Task D1.6 Inspect, test, repair, and replace electric actuator motors, relays/modules, switches, sensors, wiring, and protection devices.

Some HVAC systems control blend air and mode doors with electronic actuators. Electronic actuators contain a small motor, a gear train, and feedback device to indicate the position of the controlled door to the controlling processor. Some electronic systems contain self-diagnostic abilities with diagnostic trouble codes (DTC). You typically diagnose these with a scan tool or laptop computer. On some systems, you can adjust the actuator doors.

Task D1.7 Inspect, test, repair, or replace HVAC system electrical control panel assemblies.

Electrical control panel assemblies for manual and semi-automatic HVAC systems are modular in design, allowing for replacement of individual switches and illumination bulbs without replacing the entire panel. Most electronic ATC control panels allow the technician access only to replace illumination bulbs.

2. Air/Vacuum/Mechanical (2 questions)

Task D2.1 Diagnose the cause of failures in HVAC air, vacuum, and mechanical switches and controls; determine needed repairs.

Temperature control valves and doors that direct air flow may be controlled by electric motors, vacuum motors, or air cylinders. When air cylinders are used, they are exposed to moisture and other contaminants from the truck's air system. Often they may be restored to satisfactory service by cleaning and oiling the cylinder. One truck OEM uses an air controlled water valve in the sleeper compartment, which sometimes fails resulting in air pressure leaking into the trucks cooling system.

The most common causes of failures in HVAC air and vacuum systems are leaking hoses and diaphragms. Air and vacuum leaks can often be located by listening for a hissing noise. To locate minor air leaks, brush a mild soap solution over fittings and connections and watch for bubbles. To check for vacuum leaks, use a hand-held vacuum pump to supply 20 in. Hg to one end of the vacuum hose while the other end is plugged or attached to its device. The hose should hold 15 to 20 in. Hg. vacuum without leaking.

Task D2.2 Inspect, test, repair, or replace HVAC system air/vacuum/mechanical control panel assemblies.

HVAC vacuum and mechanical control panel assemblies require very little testing and maintenance. A leaking vacuum switch will hiss in one or more positions. Broken control cable arms or anchors will result in ineffective control levers.

Task D2.3 Inspect, test, adjust, or replace HVAC system air/vacuum/mechanical control cables and linkages.

Replace HVAC control cables if they are kinked or seized due to internal corrosion. Adjust control cables to allow a full range of motion for the control lever and for the output device.

Note: If poor A/C performance while testing is evident, a good first choice in diagnostic steps would be to ascertain the proper positioning and action of cable controls.

Task D2.4 Inspect, test, and replace HVAC system vacuum pumps, actuators (diaphragms/motors) and hoses.

Evaluate the performance of vacuum actuators and hoses by using a hand-held vacuum pump. Vacuum systems are tested by applying vacuum to the upstream (input) end of the system while individual components must be tested at the component connection. Connect the vacuum pump to each vacuum actuator and supply 15 to 20 in. Hg to the actuator. Check the vacuum actuator rod to be sure it moves freely. Close the vacuum pump valve and observe the vacuum gauge. The gauge reading should remain steady for at least one minute. If the gauge reading drops slowly, the actuator is leaking. Replace any hoses or components that do not hold vacuum or do not operate properly.

Task D2.5 Identify, inspect, test, and replace HVAC system vacuum reservoir(s), check valve(s), and restrictors.

In HVAC systems that use vacuum switches and actuators, a vacuum reservoir and check valve are installed between the vacuum source and the control panel. The check valve is a one-way valve that allows the reservoir to hold constant vacuum regardless of fluctuations in the vacuum source. The vacuum reservoir supplies vacuum at a consistent level during periods of low-source vacuum on a gasoline engine (during long uphill runs or engine lugging). The check valve must be replaced if it leaks or if it passes vacuum in both directions. The reservoir must be replaced if it will not hold vacuum. When vacuum is supplied with vacuum pump to the reservoir, it should hold 15 to 20 in. Hg.

Task D2.6 Inspect, test, adjust, repair, or replace HVAC system ducts, doors, and outlets.

Misaligned or improperly installed HVAC ducts will cause reduced levels of system output air. Blend air and mode control doors must be adjusted to allow a full range of motion when controls are operated.

To test the operation of the ducts, doors, and outlets, turn the blower to high and then move the selector through its full range. There is a mode door which changes the air flow source and a blend door that changes the temperature of the air flow. You should be able to hear doors open and close when you move the selectors. The output from the fan should change to the appropriate outlet soon after you hear the door open or close. If this does not happen, try to move the affected door with your hand. It may have something blocking it, such as small objects that may fall down the defroster outlets and become lodged in the lower ducts. The servo or actuating cylinder which is responsible for moving the door may also be stuck and need some lubrication. Replace any faulty actuators and retest the system. The appropriate service manual will show which position each door should be in the various modes. When the temperature control is changed from hot to cold, the outlet temperature should also change.

3. Constant/Automatic Temperature Control (ATC) (1 question)

Task D3.1 Diagnose constant/automatic temperature control system problems; determine needed repairs.

Most ATC systems provide internal diagnostic capabilities. On some ATC systems, diagnostic trouble codes may be displayed digitally on the control panel while on other systems a hand-held scan tool or PC interface must be used to retrieve codes. The technician should always refer to diagnostic routines in the vehicle service manual when attempting to troubleshoot ATC codes. Most importantly, the technician must rule out the possibility of mechanical failures before searching for electronic malfunctions.

Task D3.2 Inspect, test, and replace climate control temperature sensors.

ATC systems rely on a variety of sensors to provide feedback to the ATC control unit. The control unit uses the sensor signals to determine how much heating or cooling is required to maintain the desired cab or sleeper temperature. The ambient temperature sensor monitors outside air temperature. The engine coolant temperature sensor monitors engine coolant temperature. The ATC temperature (or interior temperature) sensor monitors the temperature of the air in the cab or the sleeper systems, which automatically control the temperature in the sleeper, may be designed to minimize engine idle time.

Rather than have the engine idling during driver off-duty periods, they may shut down the engine after a predetermined idle period and then automatically restart the engine when needed in response to temperature changes in the sleeper. The parameters for idle shutdown and the high and low temperature set points may be adjusted using the appropriate electronic service tools (EST).

Task D3.3 Inspect, test, repair, and replace temperature blend door/power servo system.

Most ATC systems manage temperature by blending air that has passed through the A/C evaporator core with air that has passed through the heater core. The volume of air that is allowed to pass through each core is regulated by a blend air door. The blend air door is controlled by the blend door actuator, which consists of an electric motor, a gear train, and a feedback device. The feedback device provides precise information about the position of the blend air door to the control unit. The blend door actuator must be replaced if the drive gears are worn or damaged, if the motor develops a dead spot or if the feedback device fails. A defective feedback device will cause the motor to either "hunt" for the desired position or to be inoperative.

Task D3.4 Inspect, test, adjust, or replace heater water valve and controls.

Some automatic temperature control systems use a water valve to control water flow through the heater core. They are prone to a number of problems. The valves may become clogged with deposits and seize. They should also be checked for flow restrictions. Replace them if they do not perform. Cable operated valves can suffer from bent, kinked, or misadjusted cables. Air pressure operated water valves may leak air pressure into the coolant.

Task D3.5 Inspect, test, and replace electric, air, and vacuum motors, solenoids, and switches.

In most ATC systems, the mode doors are controlled using electronic actuators that are similar to the blend door actuator. Faulty mode door actuators are usually diagnosed by following diagnostic routines in the vehicle service manual.

Task D3.6 Inspect, test, and replace constant/automatic temperature control panels.

The only serviceable components in most ATC control panels are illumination bulbs. Faulty ATC control panels are generally diagnosed by following diagnostic routines in the vehicle service manual, or with an electronic service tool.

Task D3.7 Inspect, test, and replace constant/automatic temperature control microprocessor (climate control computer/programmer).

The ATC control unit may be integral to the control panel or it may be a stand-alone component. In either event, the control unit is generally multiplexed to the engine and/or body control units, thereby providing electronic diagnostic capabilities in case of control unit failure.

E. Refrigerant Recovery, Recycling, Handling, and Retrofit (4 questions)

Note: Tasks 1 through 5 should be accomplished in accordance with published EPA and appropriate SAE "J" standards for R-12, R-134a, and EPA approved refrigerant blends.

Task E1 Maintain and verify correct operation of certified equipment.

The Clean Air Act (CAA) establishes the following rules for record keeping and operation of certified refrigeration service equipment:

1. Any person who owns approved refrigerant recycling equipment certified under the act must maintain records of the name and address of any facility to which refrigerant is sent.
2. Any person who owns approved refrigerant recycling equipment must retain records demonstrating that all persons authorized to operate the equipment are certified under the act.
3. Public Notification: Any person who conducts any retail sales of a Class I or Class II substance must prominently display a sign that reads: "It is a violation of federal law to sell containers of Class I and Class II refrigerant of less than 20 pounds of such refrigerant to anyone who is not properly trained and certified."
4. Any person who sells or distributes any Class I or Class II substance that is in a container of less than 20 pounds of such refrigerant must verify that the purchaser is certified, and must retain records for a period of three years. (These records must be maintained on-site.)

Task E2 Identify and recover A/C system refrigerant.

According to Department of Transportation (DOT)/Air Conditioning and Refrigeration Institute (ARI) guidelines, 4B4 cylinders used to store recovered refrigerant shall be painted gray with the top shoulder portion painted yellow. The refrigerant type to be stored in a given container must be clearly marked on the container's label. For recovery/recycling purposes, only cylinders that are identified for recovered refrigerant may be used. Never use a cylinder that is intended to contain new refrigerant to store recovered refrigerant. Returnable/reusable cylinders meet DOT specification 4BA-300. These cylinders are characterized by a combined liquid/vapor valve located on top.

Task E3 Recycle refrigerant.

Differences between the terms recover, recycle, and reclaim must be completely understood and properly used within the industry. To recover refrigerant is to remove refrigerant in any condition from a system and store it in an external container. The refrigerant must then either be recycled on-site or shipped off-site for reclamation. To recycle refrigerant is to reduce contaminants in used refrigerant by oil separation with single or multiple passes through devices such as replaceable filter driers, which reduce moisture, acidity, and particulate matter. To reclaim refrigerant is to reprocess refrigerant to new product specifications by means which may include distillation. Chemical analysis of the refrigerant is required to assure that appropriate product specifications are met. Reclamation usually implies the use of procedures available only at processing or manufacturing facilities.

Task E4 Handle, label and store refrigerant.

Any portable container used for transfer of reclaimed or recycled refrigerant must conform to DOT and Underwriters Laboratories (UL) standards. Before introducing refrigerant into an approved storage cylinder, the cylinder must be evacuated to at least 27 in. Hg. Cylinder-safe filling level must be monitored by measured weight. Shut-off valves are required within 12 inches (30 cm) of service hose ends. Shut-off valves must remain closed while connecting and disconnecting hoses to vehicle air conditioning service ports. Safety goggles should always be worn while working with or around refrigerant.

Task E5 Test recycled refrigerant for non-condensable gases.

To test a refrigerant for noncondensable gases, compare the pressure of the refrigerant in a cylinder to the theoretical pressure of pure refrigerant at a given temperature. If the actual pressure in the cylinder is higher than the theoretical pressure, the refrigerant is contaminated with noncondensable gas.

Pressure-Temperature Relationship

Temperature °F (°C)	R-12 PSIG (bar/kg/cm2)		R-134A PSIG (bar/kg/cm2)	
−15 (−26.1)	2.5	(.17/.18)	0	(0)
−10 (−23.3)	4.5	(.31/.32)	2.0	(.14/.14)
−5 (−20.5)	6.7	(.46/.03)	4.1	(.28/.29)
0 (−17.8)	9.2	(.63/.65)	6.5	(.45/.46)
5 (−15.0)	11.8	(.81/.83)	9.1	(.63/.64)
10 (−12.2)	14.7	(1.0/1.0)	12.0	(.89/.84)
15 (−9.4)	17.7	(1.2/1.2)	15.1	(1.0/1.2)
20 (−6.7)	21.1	(1.5/1.5)	18.4	(1.3/1.3)
25 (−3.9)	24.6	(1.7/1.7)	22.1	(1.5/1.6)
30 (−1.1)	28.5	(2.0./2.0)	26.1	(1.8/1.8)
35 (1.7)	32.6	(2.2/2.3)	30.4	(2.1/2.1)
40 (4.4)	37.0	(2.6/2.6)	35.0	(2.4/2.5)
45 (7.2)	41.7	(2.9/3.0)	40.0	(2.8/2.8)
50 (10.0)	46.7	(3.2/3.3)	45.4	(3.1/3.2)
55 (12.8)	52.1	(3.6/3.7)	51.2	(3.5/3.6)
60 (15.6)	57.8	(4.0/4.1)	57.4	(4.0/4.0)
65 (18.3)	63.8	(4.4/4.5)	64.0	(4.4/4.5)
70 (21.1)	70.2	(4.8/5.0)	71.1	(5.0/5.0)
75 (23.9)	77.0	(5.3/5.4)	78.6	(5.4/5.5)
80 (26.7)	84.2	(5.8/6.0)	86.7	(6.0/6.1)
85 (29.4)	91.7	(6.3/6.4)	95.2	(6.6/6.7)
90 (32.2)	99.7	6.9/7.0)	104.3	(7.2/7.3)
95 (35.0)	108.2	(7.5/7.6)	113.9	(7.9/8.0)
100 (37.8)	117.0	(8.1/8.2	124.1	(8.6/8.7)
105 (40.6)	126.4	(8.7/8.9)	134.9	(9.3/9.5)
110 (43.3)	136.2	(9.4/9.6)	146.3	(10.1/10.3)
115 (46.1)	146.5	(10.1/10.3)	158.4	(11.0/11.1)
120 (48.9)	157.3	(11.0/11.1)	171.1	(11.8/12.0)

Task E6 Retrofit A/C systems following industry/manufacturers' accepted procedures and federal/local laws.

When retrofitting systems to use R-134a refrigerant, it is recommended that procedures be followed as outlined by OEMs and be in compliance with all federal and local statutes. This includes recovery, evacuation, and replacement of components.

5 Sample Test for Practice

Sample Test

Please note the letter and number in parentheses following each question. They match the task in Section 4 that discusses the relevant subject matter. You may want to refer to the overview using the cross-referencing key to help with questions posing problems for you.

1. Which of the following should be used to lubricate replacement A/C hose O-rings?
   A. petroleum jelly
   B. transmission fluid
   C. silicone grease
   D. refrigeration oil (B3.2)

2. Technician A says the R-12 A/C service valve has a removable core. Technician B says the R-134a service valve must be rear-seated during A/C compressor replacement. Who is correct?
   A. A only
   B. B only
   C. Both A and B
   D. Neither A nor B (B3.10)

3. Refrigerant oil must be added to the new component prior to installation when replacing any of the following EXCEPT
   A. the evaporator.
   B. the condenser.
   C. the accumulator.
   D. the TXV. (B3.1)

4. Which of the following tools can be used to measure the freeze protection of engine coolant?
   A. litmus strip
   B. manometer
   C. refractometer
   D. opacity meter (C3)

5. Manual temperature control systems may use a coolant control valve that is operated using any of the following EXCEPT
   A. chassis air pressure.
   B. vacuum.
   C. a cable.
   D. a magneto. (D3.4)

6. The blend door actuator motor is usually mounted
   A. using epoxy.
   B. using butylene sealer.
   C. behind the glove box.
   D. to the evaporator case. (D1.6)

7. An electric cooling fan motor can be controlled by any of the following EXCEPT
   A. an electronic relay.
   B. an independent electronic module.
   C. a multiplexed electronic module.
   D. an air solenoid controller. (D1.5)

8. All of the following can cause A/C compressor drive belt misalignment EXCEPT
   A. a bent compressor mounting bracket holder.
   B. a faulty compressor clutch bearing assembly.
   C. an improperly set compressor clutch air gap.
   D. a faulty idler pulley. (B2.3)

9. Before a portable container is used to transfer recycled R-12, it must be evacuated to at least
   A. 20 in. Hg.
   B. 22 in. Hg.
   C. 27 in. Hg.
   D. 12 in. Hg. (E4)

10. In servicing the expansion valve, which of the following can a technician perform?
    A. Adjust the expansion valve using a torque wrench.
    B. Adjust the expansion valve using an Allen wrench.
    C. Adjust the expansion valve using a screwdriver.
    D. The expansion valve cannot be adjusted. (B3.6)

11. To avoid mixing refrigerants, Technician A says that the R-134a service hose fittings are different from R-12 fittings. Technician B says that R-12 containers may be either white or yellow, but R-134a containers are sky blue. Who is correct?
    A. A only
    B. B only
    C. Both A and B
    D. Neither A nor B (E2)

12. Technician A says the HVAC control panel sometimes contains replaceable components. Technician B says that you must replace the HVAC control panel as a unit, if any of the components fail. Who is correct?
    A. A only
    B. B only
    C. Both A and B
    D. Neither A nor B (D1.7)

13. The lubricant used in R-134a mobile A/C systems is
    A. a PAG-based lubricant.
    B. a mineral-based petroleum lubricant.
    C. a DEXRON or DEXRON II lubricant.
    D. a type C-3 SAE 30 based oil lubricant. (B1.9)

14. Technician A says that the heater control valve can be operated by engine speed. Technician B says that the heater control valve can be cable-operated. Who is correct?
    A. A only
    B. B only
    C. Both A and B
    D. Neither A nor B (C10)

15. The process that reduces contaminants in used refrigerant by using oil separation and filter core dryers is
    A. restoration.
    B. recovery.
    C. recycling.
    D. reclamation. (E3)

16. A heater does not supply the cab with enough heat. The coolant level and blower test are OK. Technician A says an improperly adjusted temperature control cable could be the cause. Technician B says a clogged heater core could be the cause. Who is correct?
    A. A only
    B. B only
    C. Both A and B
    D. Neither A nor B (C1)

17. The A/C high pressure relief valve shows evidence of slight oil leakage. Technician A says you must replace the valve and repair the leak. Technician B says that if the valve is just leaking a little oil and not refrigerant, repair is not necessary. Who is correct?
    A. A only
    B. B only
    C. Both A and B
    D. Neither A nor B (B3.11)

18. The low pressure cut-out switch senses pressure in the
    A. system high side.
    B. atmosphere.
    C. system low side.
    D. cab/sleeper. (B2.2)

19. All of the following are good methods of verifying that an engine thermostat opens EXCEPT
    A. feeling the upper radiator hose.
    B. watching the temperature gauge.
    C. watching for motion in the upper radiator tank.
    D. watching the surge tank. (C7)

20. Technician A says that epoxy should be used to repair a broken temperature control lever. Technician B says that a temperature control head with a broken lever should be replaced. Who is correct?
    A. A only
    B. B only
    C. Both A and B
    D. Neither A nor B (D2.2)

21. When evacuating an A/C system, which manifold gauge hose is connected to the vacuum pump?
    A. the high pressure hose
    B. the low pressure hose
    C. the center service hose
    D. any hose (B1.6)

22. What purpose does a vacuum check valve have in an HVAC system?
    A. to monitor vacuum in all HVAC systems
    B. to delay vacuum to the actuators until the coolant has reached operating temperature
    C. to prevent shock damage to vacuum actuator diaphragms
    D. to prevent loss of vacuum to components during periods of low engine vacuum (D2.5)

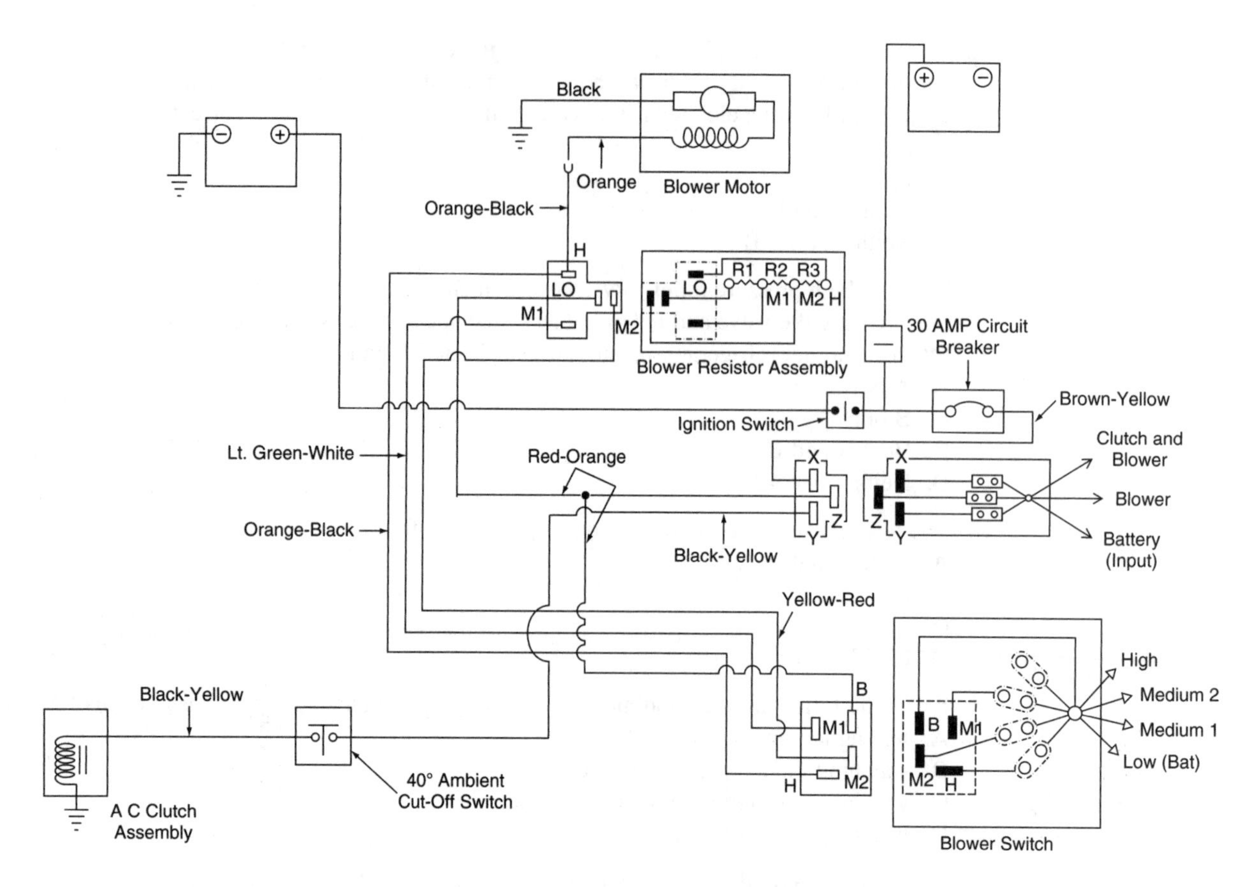

23. Refer to the figure above. All of the following could prevent the A/C clutch from engaging EXCEPT
    A. a faulty 30-amp circuit breaker.
    B. ambient temperature below 40°F.
    C. the ambient temperature cutout switch stuck closed.
    D. open circuit in the black/yellow wire. (D1.1)

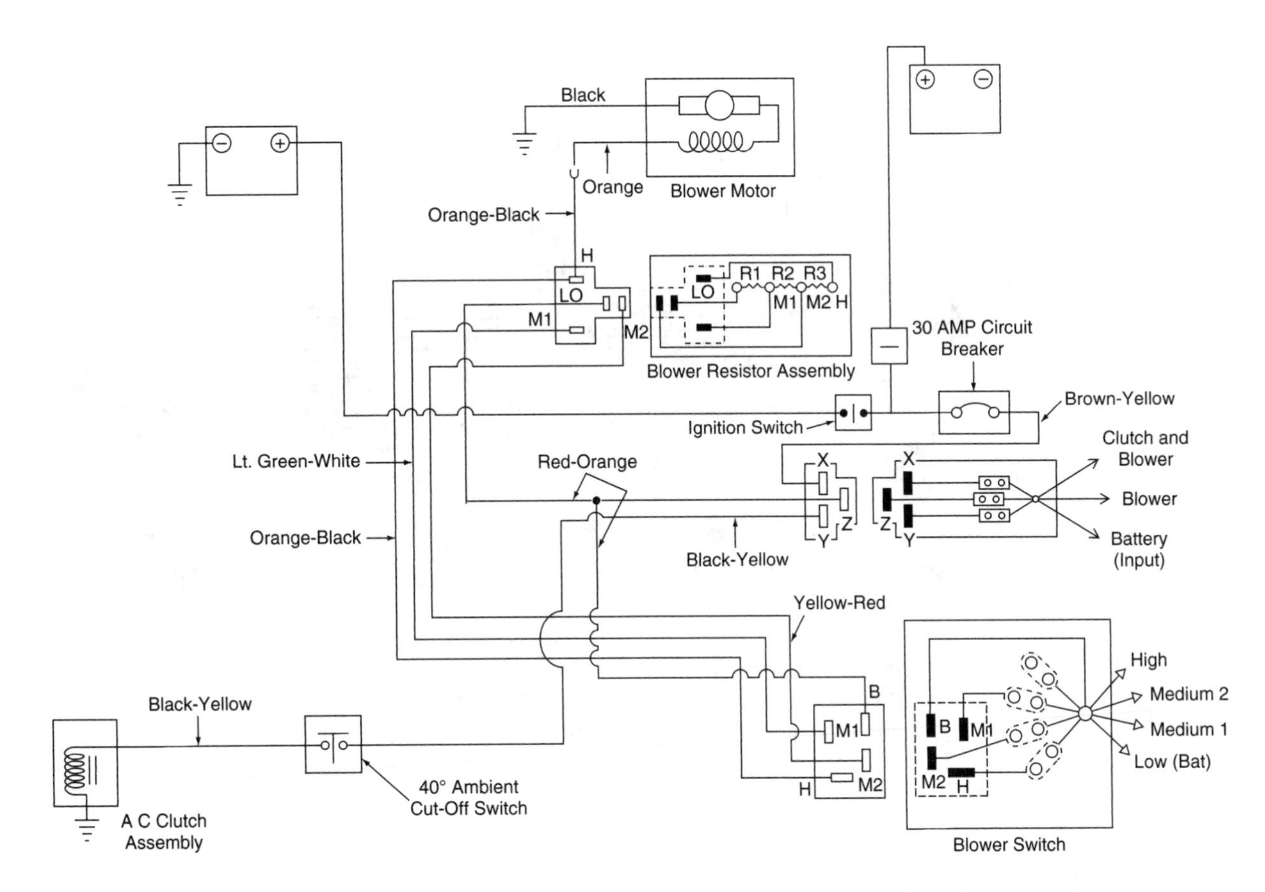

24. Refer to the figure above. When the blower switch is in the medium-2 position, how many resistors are used in the circuit to control blower motor speed?
A. three
B. two
C. one
D. none (D1.2)

25. Technician A says that if there is no output from the blower motor switch, you replace the switch. Technician B says that feedback from the blower motor resistors could cause the switch to malfunction. Who is correct?
A. A only
B. B only
C. Both A and B
D. Neither A nor B (D2.1)

26. When attempting to verify a leaking evaporator core, the technician is LEAST-Likely to sense refrigerant with the detector probe
A. at the evaporator case drain.
B. over the block type expansion valve.
C. at the panel vents.
D. at the defroster vents. (B3.8)

27. Technician A says that a cracked fan blade should be welded. Technician B says that a cracked fan blade can be repaired with epoxy. Who is correct?
A. A only
B. B only
C. Both A and B
D. Neither A nor B (C9)

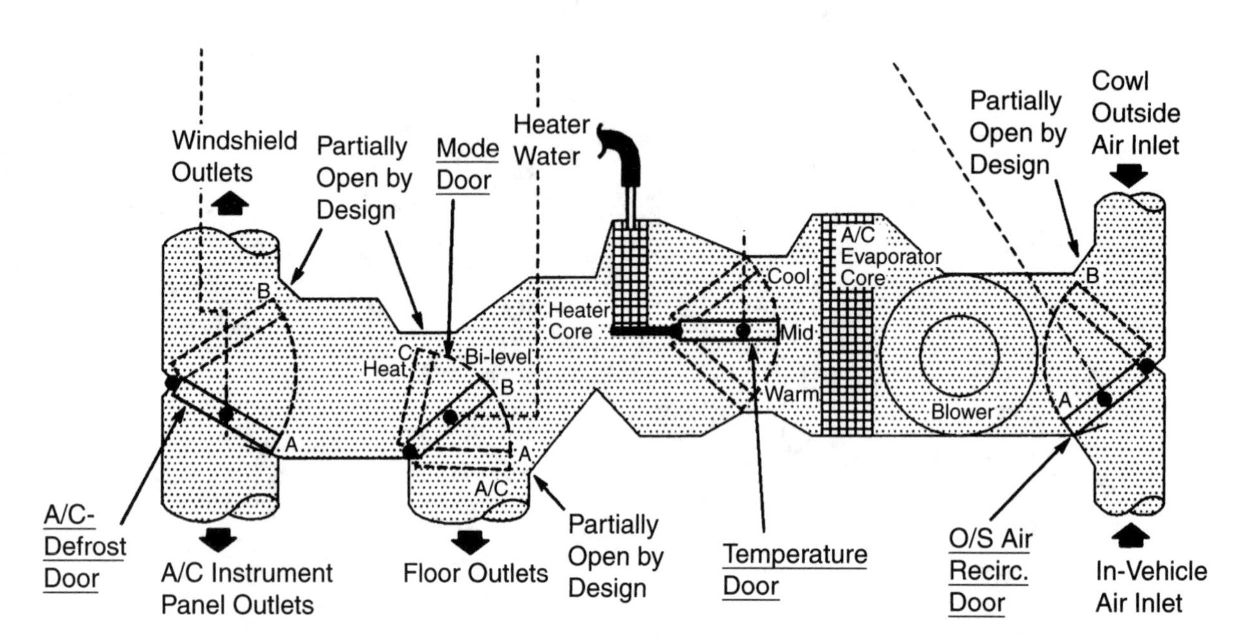

28. The outside air recirculation door is stuck in position A in the figure above. Technician A says under this condition outside air is drawn into the HVAC inlet. Technician B says under this condition in-vehicle air is recirculated. Who is correct?
    A. A only
    B. B only
    C. Both A and B
    D. Neither A nor B (D2.6, D2.7)

29. One segment of the digital readout on an ATC control panel is inoperative. Technician A says that the control panel should be replaced. Technician B says that the Light Emitting Diode (LED) can be replaced. Who is correct?
    A. A only
    B. B only
    C. Both A and B
    D. Neither A nor B (D3.6)

30. The upper radiator hose has a slight bulge. Technician A says that the hose does not need to be replaced unless it appears cracked. Technician B says that the bulge indicates a weak spot and the hose should be replaced. Who is correct?
    A. A only
    B. B only
    C. Both A and B
    D. Neither A nor B (C4)

31. Owners of approved refrigerant recycling equipment must maintain records that demonstrate
    A. only certified technicians operate the equipment.
    B. the equipment is operated under the supervision of certified technicians.
    C. technicians who are undergoing certification training operate the equipment.
    D. the equipment is operated only when a certified technician is on the premises. (E1)

32. Technician A says that the binary pressure switch prevents compressor operation if the refrigerant charge has been lost or ambient temperature too cold. Technician B says that the binary pressure switch turns off the compressor if the system pressure is too high. Who is correct?
    A. A only
    B. B only
    C. Both A and B
    D. Neither A nor B (D1.3)

33. All of the following statements about air-controlled HVAC systems in heavy-duty trucks are true EXCEPT
    A. air cylinders are used to open shutters.
    B. coolant control valves can be controlled using chassis air.
    C. air cylinders are used to control mode and blend air doors.
    D. air leaks may cause mode doors to move slowly or to be totally inoperative. (D2.3)

34. Technician A says that the refrigerant containers for R-12 and R-134a are color coded. Technician B says that the R-134a containers use ½ inch, 16 ACME threads that cannot be mistakenly connected to an R-12 gauge set or recovery station. Who is correct?
    A. A only
    B. B only
    C. Both A and B
    D. Neither A nor B (B1.2)

35. During a performance test, the A/C compressor clutch is observed to be slipping. Technician A says that the air gap was probably set improperly. Technician B says that the pressure plate needs to be resurfaced. Who is correct?
    A. A only
    B. B only
    C. Both A and B
    D. Neither A nor B (B2.4)

36. Air flow through the A/C condenser will be significantly affected by all of the following EXCEPT
    A. debris trapped in the fins.
    B. relative humidity of the outside air.
    C. bent cooling fins.
    D. vehicle speed. (B3.3)

37. While using an electronic leak detector, where should the detector's probe be placed near a suspected leak in order to most effectively find the leak?
    A. directly above the suspected leak area
    B. directly below the suspected leak area
    C. 6 inches upstream from the suspected leak area
    D. 6 inches downstream from the suspected leak area (B1.5)

38. Technician A says that some ATC control panels can be recalibrated using a handheld digital diagnostic tool. Technician B says that a handheld digital diagnostic tool can be used to help diagnose ATC system failures. Who is correct?
    A. A only
    B. B only
    C. Both A and B
    D. Neither A nor B (D3.1, D3.6, D3.7)

39. The best instrument to use when troubleshooting an A/C electronic circuit with solid-state components is
    A. a DMM.
    B. a self-powered test lamp.
    C. an analog volt/ohmmeter.
    D. a 12-volt test lamp. (D3.7)

40. A growling noise is coming from the water pump. Technician A says that the water pump bearing could be the cause. Technician B says that the impeller has been eroded by cavitation. Who is correct?
    A. A only
    B. B only
    C. Both A and B
    D. Neither A nor B (C6)

41. A cracked A/C compressor mounting plate could cause all of the following symptoms EXCEPT
    A. drive belt wear.
    B. internal compressor damage.
    C. vibration with the A/C compressor clutch engaged.
    D. drive belt squeal or chatter. (B2.7)

42. All of these statements about cooling system service are true EXCEPT
    A. when the cooling system pressure is increased, the boiling point is decreased.
    B. if more antifreeze is added to the coolant mix, the boiling point is increased.
    C. a good quality ethylene glycol antifreeze contains a corrosion inhibitor.
    D. coolant solutions must be recovered, recycled, or handled as hazardous material. (C8)

43. The proper method of testing a vacuum actuator is to
    A. apply shop air to the vacuum port and listen for leaks.
    B. use a handheld vacuum pump and observe the gauge.
    C. move the control lever and verify that the actuator plunger moves.
    D. apply vacuum using an A/C evacuation pump and verify that the actuator plunger moves. (D2.4)

44. With the HVAC system in DEFROST mode, the blower on HIGH, and the temperature control on COLD, the bottom of the windshield fogs up on the outside because
    A. the evaporator case drain is clogged.
    B. the heater core leaks.
    C. the evaporator core is iced up.
    D. the cold windshield causes moisture to condense from the outside air. (C2)

45. While testing an A/C system at normal speed and temperature, the high-side and low-side pressures are the same. Technician A says the clutch may not be engaged. Technician B says there may be a restriction in the expansion valve. Who is correct?
    A. A only
    B. B only
    C. Both A and B
    D. Neither A nor B (B1.3)

46. While testing a heater core for leaks, you apply 10 psi of air pressure. The pressure bleeds to 5 psi in 3 minutes. You should conclude that
    A. the core is OK.
    B. the core may require repair in future.
    C. the core leaks and needs repair now.
    D. this is not a valid test for a leaking heater core. (C11)

47. Technician A says that whenever the A/C system has been open to atmosphere for an extended time, the receiver/drier should be replaced. Technician B says that when a compressor is replaced, the receiver drier should also be replaced. Who is correct?
    A. A only
    B. B only
    C. Both A and B
    D. Neither A nor B (B3.5)

48. Raising the pressure in the cooling system
    A. lowers the boiling point of the coolant.
    B. raises the boiling point of the coolant.
    C. prevents corrosion in the cooling system.
    D. does not affect the boiling point of the coolant. (C5)

49. A coolant temperature sensor is classified as NTC (negative temperature coefficient). Technician A says this is a thermistor in which internal resistance decreases in proportion to temperature rise. Technician B says that in most truck engine cooling systems, a thermistor is supplied with V-Bat (battery voltage) and returns a portion of it as a signal. Who is correct?
    A. A only
    B. B only
    C. Both A and B
    D. Neither A nor B (D1.4, D3.2)

50. Technician A says that deformed or improperly aligned condenser mounting insulators will damage the A/C compressor. Technician B says that deformed or improperly aligned condenser mounting insulators could damage the condenser and refrigerant lines. Who is correct?
    A. A only
    B. B only
    C. Both A and B
    D. Neither A nor B (B3.4)

51. A sweet odor comes from the panel vents while operating the A/C in NORMAL mode. Technician A says he can smell R-12 in the cab, so the truck must have a leaking evaporator. Technician B says the heater core could be leaking. Who is correct?
    A. A only
    B. B only
    C. Both A and B
    D. Neither A nor B (A3)

52. ATC systems may use any of the following sensors EXCEPT
    A. a sunlight sensor.
    B. an ambient temperature sensor.
    C. an evaporator temperature sensor.
    D. a manifold pressure sensor. (D3.2)

53. A truck's A/C system does not cool when driving at slow speeds. Technician A states that the truck needs a fan clutch. Technician B states that the truck needs to be road tested to see if the A/C cools well at highway speeds. Who is correct?
    A. A only
    B. B only
    C. Both A and B
    D. Neither A nor B (A1)

54. An HVAC system outputs air at a constant temperature, regardless of the temperature setting. The most likely cause is
    A. a low refrigerant charge.
    B. low coolant level.
    C. a compressor clutch failure.
    D. a broken blend door cable. (A4, B1.1)

55. To remove a mildew odor from the A/C output air
    A. remove the evaporator case, clean it with a vinegar and water solution, and dry it thoroughly before reinstallation.
    B. spray a disinfectant into the panel outlets.
    C. pour a small amount of alcohol into the air intake plenum.
    D. place an automotive deodorizer under the dash. (B3.9)

56. During normal A/C operation a loud hissing noise is heard and a cloud of vapor is discharged from under the vehicle. Technician A says that excessive high-side pressure that caused the pressure relief valve on the A/C compressor to trip may have been the result of a defective engine cooling fan clutch. Technician B says that the relief valve may have tripped due to a defective shutter control solenoid. Who is correct?
    A. A only
    B. B only
    C. Both A and B
    D. Neither A nor B (B2.1, B3.11, C9)

57. What evidence should a technician look for to determine if a truck has had a retrofit from R-12 to R-134a?
    A. The compressor has been replaced.
    B. The dryer has been replaced.
    C. The caps are missing from the service ports.
    D. A retrofit decal is installed on the fire wall. (E6)

58. A whistling noise coming from under the dash while the HVAC system is being operated with the blower on HIGH could indicate
    A. a misaligned duct.
    B. a defective vacuum actuator.
    C. an improperly adjusted mode door cable.
    D. a poor electrical connection to the blend door motor. (D2.6)

59. The recommended tool for diagnosing electronic sensors and actuators in an ATC system is
    A. a 12-volt test light.
    B. an analog multimeter.
    C. a digital multimeter (DMM).
    D. an A/C system charging station. (D3.5)

Pressure-Temperature Relationship

Temperature ° F (°C)	R-12 PSIG	(bar/kg/cm2)	R-134A PSIG	(bar/kg/cm2)
–15 (–26.1)	2.5	(.17/.18)	0	(0)
–10 (–23.3)	4.5	(.31/.32)	2.0	(.14/.14)
–5 (–20.5)	6.7	(.46/.03)	4.1	(.28/.29)
0 (–17.8)	9.2	(.63/.65)	6.5	(.45/.46)
5 (–15.0)	11.8	(.81/.83)	9.1	(.63/.64)
10 (–12.2)	14.7	(1.0/1.0)	12.0	(.89/.84)
15 (–9.4)	17.7	(1.2/1.2)	15.1	(1.0/1.2)
20 (–6.7)	21.1	(1.5/1.5)	18.4	(1.3/1.3)
25 (–3.9)	24.6	(1.7/1.7)	22.1	(1.5/1.6)
30 (–1.1)	28.5	(2.0./2.0)	26.1	(1.8/1.8)
35 (1.7)	32.6	(2.2/2.3)	30.4	(2.1/2.1)
40 (4.4)	37.0	(2.6/2.6)	35.0	(2.4/2.5)
45 (7.2)	41.7	(2.9/3.0)	40.0	(2.8/2.8)
50 (10.0)	46.7	(3.2/3.3)	45.4	(3.1/3.2)
55 (12.8)	52.1	(3.6/3.7)	51.2	(3.5/3.6)
60 (15.6)	57.8	(4.0/4.1)	57.4	(4.0/4.0)
65 (18.3)	63.8	(4.4/4.5)	64.0	(4.4/4.5)
70 (21.1)	70.2	(4.8/5.0)	71.1	(5.0/5.0)
75 (23.9)	77.0	(5.3/5.4)	78.6	(5.4/5.5)
80 (26.7)	84.2	(5.8/6.0)	86.7	(6.0/6.1)
85 (29.4)	91.7	(6.3/6.4)	95.2	(6.6/6.7)
90 (32.2)	99.7	6.9/7.0)	104.3	(7.2/7.3)
95 (35.0)	108.2	(7.5/7.6)	113.9	(7.9/8.0)
100 (37.8)	117.0	(8.1/8.2	124.1	(8.6/8.7)
105 (40.6)	126.4	(8.7/8.9)	134.9	(9.3/9.5)
110 (43.3)	136.2	(9.4/9.6)	146.3	(10.1/10.3)
115 (46.1)	146.5	(10.1/10.3)	158.4	(11.0/11.1)
120 (48.9)	157.3	(11.0/11.1)	171.1	(11.8/12.0)

60. A storage container or air conditioning system containing R-12 (at rest) and subject to an ambient temperature of 70°F will have an internal gauge pressure of approximately (see table)
    A. 220 psi.
    B. 125 psi.
    C. 70 psi.
    D. 30 psi. (E5)

61. All of the following statements about charging an A/C system are true EXCEPT
    A. refrigerant may be installed through both service valves when the engine is not running.
    B. refrigerant may be installed through the low-side service valve when the engine is running.
    C. refrigerant may be installed through both service ports when the engine is running.
    D. you may install refrigerant directly from an approved charging station. (B1.8)

62. The refrigerant in an A/C system starts changing state from liquid to vapor as it flows through the
    A. vapor line.
    B. condenser.
    C. fixed orifice.
    D. capillary tube. (B3.7)

63. On a vehicle equipped with ATC, the blend door actuator motor runs when the temperature setting is changed, but the blend door does not move. The most likely cause of this problem is
    A. a defective control module.
    B. a defective actuator feedback device.
    C. a defective drive gear in the actuator.
    D. an improperly adjusted ATC sensor cable. (D3.3)

64. A customer complains that he is unable to change the output temperature of his HVAC system. The most likely cause of this problem is
    A. a broken blend door cable.
    B. a defective compressor clutch.
    C. a clogged orifice tube.
    D. a defective blower switch. (B1.1)

65. A band of frost on the A/C high pressure hose upstream from the orifice tube indicates
    A. a defective compressor discharge valve.
    B. a restriction in the high pressure hose.
    C. a clogged orifice tube.
    D. moisture in the system. (B1.4)

66. There is a growling or rumbling noise at the A/C compressor with the compressor clutch engaged or disengaged. Technician A says the compressor bearing is defective. Technician B says the clutch bearing is defective. Who is correct?
    A. A only
    B. B only
    C. Both A and B
    D. Neither A nor B (A2)

67. Which is the best method of removing particulate from an A/C system after a mechanical failure?
    A. A/C flush solvent in the reverse flow
    B. R-11 flushing
    C. in-line filter
    D. nitrogen flushing (B1.7)

68. To check and adjust the A/C compressor lubricant level
    A. quickly purge the system and add oil charges to refill it.
    B. open the drain plug and crank the engine until the compressor is empty, then pump fresh oil into the compressor.
    C. remove the compressor from the vehicle, drain the oil, and add the specified quantity of fresh oil.
    D. add refrigerant oil until you can see the oil level. (B2.5)

69. Technician A says that when installing a new or rebuilt compressor you should first turn it over by hand and drain any oil shipped with the compressor, and then install the correct amount of the specified oil before installing it. Technician B says that some compressors are shipped without oil. Who is correct?
    A. A only
    B. B only
    C. Both A and B
    D. Neither A nor B (B2.6)

6 Additional Test Questions for Practice

Additional Test Questions

Please note the letter and number in parentheses following each question. They match the task in Section 4 that discusses the relevant subject matter. You may want to refer to the overview using the cross-referencing key to help with questions posing problems for you.

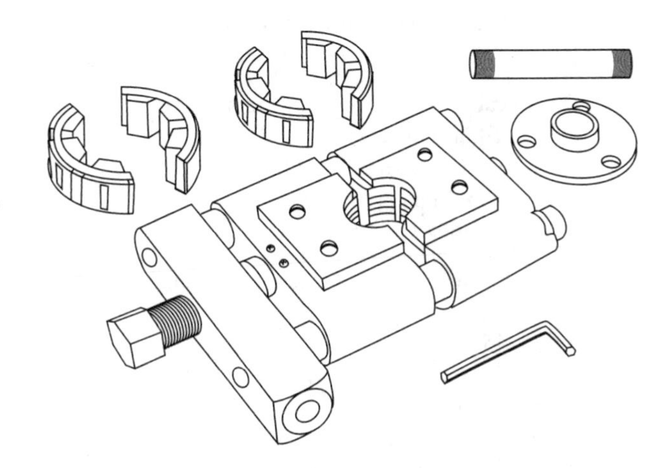

1. What is the function of the piece of equipment in the figure?
   A. spring lock installation tool
   B. bearing puller
   C. hose end crimping tool
   D. flare tool (B3.2)

2. How much refrigeration oil should be in a typical A/C condenser?
   A. one ounce
   B. five ounces
   C. seven ounces
   D. eleven ounces (B3.1)

3. The expansion valve is located at
   A. the inlet line of the evaporator.
   B. the outlet line of the evaporator.
   C. the inlet line of the compressor.
   D. the outlet line of the condenser. (B3.6)

4. A typical ATC system will delay blower motor operation until coolant temperature reaches
   A. 90°F (32°C).
   B. 70°F (21°C).
   C. 100°F (38°C).
   D. 120°F (49°C). (D3.2)

5. If R-12 comes into contact with a flame
   A. it will explode.
   B. it will form a nontoxic gas.
   C. it will form chlorine crystals.
   D. it will form phosgene gas. (B1.5)

6. A cooling fan clutch can be controlled by any of the following EXCEPT
   A. an air clutch.
   B. a thermostatic spring and fan clutch fluid.
   C. an electrically actuated clutch.
   D. a hydraulic switch. (C9)

7. Never operate the compressor with the high-side service valve
   A. back-seated.
   B. front-seated.
   C. in mid-position.
   D. hot. (B3.10)

8. Which cab heating system is the most common type used in trucks?
   A. forced air convection heater
   B. immersion heater
   C. fuel-fired heater
   D. electric heater (A4)

9. Of the following, which is a normal range of low-side gauge operating pressures?
   A. 5–10 psi
   B. 25–45 psi
   C. 60–80 psi
   D. 180–205 psi (B1.3)

10. The inside of the windshield has a sticky film. Technician A says to check the engine coolant level. Technician B says the heater core may be leaking. Who is correct?
    A. A only
    B. B only
    C. Both A and B
    D. Neither A nor B (C2)

11. Cooling and heating system hoses should be replaced for any of the following reasons EXCEPT
    A. they leak at the hose clamps.
    B. they are cracked.
    C. they show signs of bulging.
    D. they feel spongy. (C4)

12. Approximately how much refrigerant oil must be added to a newly replaced evaporator core?
    A. none
    B. 3 ounces
    C. 9 ounces
    D. 14.5 ounces (B3.8)

13. Any of the following can be used to clean road debris from the condenser fins EXCEPT
    A. a mild saline solution.
    B. a soft whisk broom.
    C. compressed air.
    D. a mild soap and water solution. (B3.3)

14. Which of these characteristics does the R-134a refrigerant possess?
    A. odorless
    B. a faint ether-like odor
    C. a strong rotten egg odor
    D. a cabbage-like odor (B1.2)

15. Which of the following items is Most-Likely to cause elevated high-side pressure in an A/C system?
    A. restricted air flow through condenser
    B. a thermostat stuck open
    C. a leaking thermal bulb
    D. an open bypass valve (B1.3)

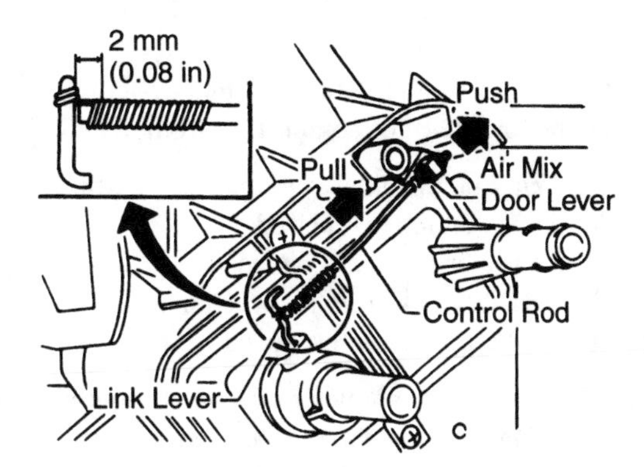

16. What adjustment is being performed in the figure above?
    A. the air mix door adjustment
    B. the manual coolant valve adjustment
    C. the ventilation door control rod adjustment
    D. the defroster door control rod adjustment (D2.6)

17. Technician A says all A/C systems use a mineral-based internal lubricant. Technician B says PAG oil is a nonsynthetic form of lubricant. Who is correct?
    A. A only
    B. B only
    C. Both A and B
    D. Neither A nor B (B1.9)

18. Which of the following is the LEAST-Likely cause of a binding temperature control cable?
    A. a kinked cable housing
    B. corrosion in the cable housing
    C. a deformed or overtightened cable clamp
    D. a defective mode door (D2.3)

19. To check for cooling system external leaks, pressurize the system and
    A. use a black light to find leaks.
    B. use a leak detector to find leaks.
    C. perform a visual inspection.
    D. watch for white smoke. (C3)

20. To detect a refrigerant leak, hold the leak detector sensor
    A. within 3 inches of the fitting.
    B. just below the fitting.
    C. right next to the fitting.
    D. just above the fitting. (B1.5)

21. An electric condenser fan may be controlled by any of the following EXCEPT
    A. the engine control unit.
    B. the chassis management unit.
    C. a manual switch.
    D. an electronic relay. (D1.5)

22. An ATC system controls all of the following EXCEPT
    A. the engine radiator shutters.
    B. the blower motor speed.
    C. the blend door actuator.
    D. the outside air door position. (D3.1)

23. Flushing the cooling system does not
    A. remove rust from the system.
    B. remove contaminants from the system.
    C. increase the life of the cooling system.
    D. remove acids of combustion from the cooling system. (C8)

24. The water pump should be replaced any time
    A. the fan clutch is replaced.
    B. the heater hoses are replaced.
    C. there is a leak from the weep hole.
    D. the thermostat sticks closed. (C6)

25. Before replacing an electric blend air door actuator, the technician should
    A. ensure that the batteries are removed from the vehicle.
    B. ensure that the batteries have a good ground.
    C. ensure that the blend door moves freely.
    D. ground himself to the vehicle. (D1.6)

26. Which of the following is a typical heavy-duty-truck cooling system operating pressure?
    A. 3 psi
    B. 15 psi
    C. 20 psi
    D. 30 psi (C5)

27. Technician A says a TXV not regulating properly could cause reduced air flow from the instrument panel outlets. Technician B says a TXV not regulating properly could cause the evaporator to ice up. Who is correct?
    A. A only
    B. B only
    C. Both A and B
    D. Neither A nor B (B1.1)

28. In MAX A/C mode
    A. the outside air door is closed.
    B. the defroster door is open.
    C. the compressor clutch cannot disengage.
    D. the blower is disabled. (B1.1)

29. If the blower operates in all speeds except medium, which of the following problems is the Most-Likely cause?
    A. a defective blower switch
    B. an intermittent short in the blower motor
    C. a defective blower relay
    D. a loose terminal in the motor connector (D1.2)

30. Technician A says that some systems have an inline filter. Technician B says that petroleum jelly should be used to lubricate new A/C O-rings. Who is correct?
    A. A only
    B. B only
    C. Both A and B
    D. Neither A nor B (B3.2)

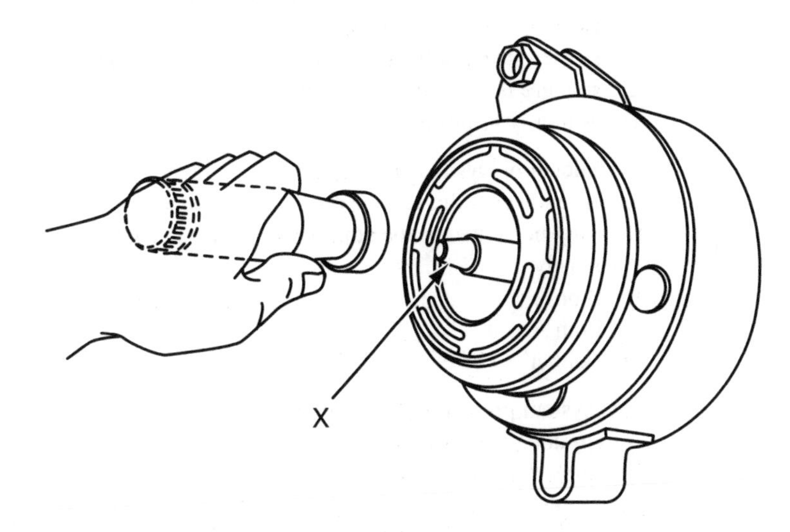

31. Technician A says component X in the figure above is a seal protector. Technician B says you must install the seal seat O-ring before the compressor shaft seal. Who is correct?
    A. A only
    B. B only
    C. Both A and B
    D. Neither A nor B (B2.7)

32. Over-tightening the mounting screws of an HVAC system vacuum actuator is Most-Likely to result in
    A. a ruptured diaphragm.
    B. stripped threads.
    C. deformed linkage.
    D. a vacuum leak. (D2.4)

33. Poor cooling from the A/C system that uses a TXV valve can be caused by all of the following EXCEPT
    A. the fan clutch always engaged.
    B. an improperly adjusted blend door cable.
    C. a low refrigerant charge.
    D. a refrigerant overcharge. (B1.1)

34. A routine A/C maintenance service should include all of the following EXCEPT
    A. tightening the condenser lines.
    B. removing debris from the condenser fins.
    C. straightening the condenser fins.
    D. checking the condenser mounts. (B3.4)

35. Technician A states that systems that use an orifice tube use an accumulator. Technician B states that systems that use a TXV use a receiver-dryer. Who is correct?
    A. A only
    B. B only
    C. Both A and B
    D. Neither A nor B (A.4, B3.6, B3.7)

36. All of these statements are true about coolant control valves EXCEPT
    A. it controls the flow of coolant through the heater core.
    B. it is part of the water pump.
    C. it may be cable operated.
    D. it may be vacuum operated. (C10)

37. The A/C compressor drive belt should be adjusted
    A. using a belt tension gauge.
    B. so there is no deflection at maximum engine speed.
    C. so there is no static deflection.
    D. as tightly as possible. (B2.3)

38. Technician A says that the easiest way to identify the type of refrigerant that a system should use is to observe the service port fittings. Technician B says that R-134a is the only refrigerant that may be vented to the atmosphere. Who is correct?
    A. A only
    B. B only
    C. Both A and B
    D. Neither A nor B (B1.2)

39. A/C system pressures vary with all of the following EXCEPT
    A. altitude.
    B. ambient temperature.
    C. ambient air humidity
    D. cab humidity. (B1.3)

40. Technician A says that some ATC systems use a microprocessor built into the control panel. Technician B says that some ATC systems use a microprocessor that can be replaced independently from the control panel. Who is correct?
    A. A only
    B. B only
    C. Both A and B
    D. Neither A nor B (D3.7)

41. Coolant conditioner performs all of the following tasks EXCEPT
    A. raises the boiling point of the coolant.
    B. inhibits rust and debris from the coolant.
    C. lubricates the cooling system internally.
    D. prevents cavitation corrosion of wet cylinder liners. (C3)

42. Technician A says that as specified by SAE J1991, recycled refrigerant cannot contain more than 15 ppm moisture contaminants by weight. Technician B says that as specified by SAE J1991, recycled refrigerant cannot contain more than 330 ppm noncondensable gases (air) by weight. Who is correct?
    A. A only
    B. B only
    C. Both A and B
    D. Neither A nor B (E3)

43. A blown HVAC system fuse could indicate any of the following EXCEPT
    A. a short circuit to ground in the blower circuit.
    B. a short circuit to ground in the blend door actuator.
    C. a short circuit in the engine coolant temperature (ECT) sensor.
    D. a damaged wiring harness connector. (D1.1)

44. When pressure testing a cooling system, there are no obvious external leaks, but the system cannot maintain pressure. The Most-Likely cause of this problem is
    A. a leaking evaporator.
    B. a defective heater valve.
    C. a stuck-open thermostat.
    D. a blown head gasket in the engine. (C3)

45. Before removing any chassis air hose, the technician must
    A. start the engine.
    B. drain water from the air system.
    C. drain all of the air from the system.
    D. remove the air compressor from the vehicle. (D2.6)

46. Which of these terms correctly describes refrigerant that has been removed from a system and stored in an external container?
    A. recycled
    B. recovered
    C. reclaimed
    D. refined (E3)

47. When replacing the heater core, foam tape is used
    A. to insulate the heater core.
    B. to reduce noise from coolant surges.
    C. to protect and seal around the heater core.
    D. to seal the heater hose connections. (C11)

48. A coiled spring inside a radiator hose is used to
    A. pre-form the hose.
    B. prevent the hose from collapsing.
    C. increase the hose burst pressure.
    D. eliminate cavitation. (C4)

49. Inadequate air flow from one vent could be caused by
    A. a misaligned air duct.
    B. a faulty blower resistor.
    C. a clogged heater core.
    D. high ambient humidity. (D2.6)

50. Technician A says that the evaporator must be removed from the vehicle if the presence of evaporator lubricant is to be checked. Technician B says that the refrigeration oil is distributed throughout the A/C system. Who is correct?
    A. A only
    B. B only
    C. Both A and B
    D. Neither A nor B (B3.1)

51. A compressor cycling on and off too fast is most commonly caused by
    A. a defective compressor clutch.
    B. a defective control switch.
    C. an overcharged system.
    D. a low refrigerant charge. (B1.4)

52. The LEAST-Likely cause of poor coolant circulation in a truck with a downflow radiator is
    A. a defective thermostat.
    B. an eroded water pump impeller.
    C. a collapsed upper radiator hose.
    D. a collapsed lower radiator hose. (C6, C7)

53. On accumulator-type systems with the compressor cycling switch located on the accumulator, the switch senses
    A. outside temperature.
    B. accumulator pressure.
    C. accumulator temperature.
    D. engine compartment temperature. (D1.3)

54. A technician finds an air leak in the HVAC control panel. The best method of repair is
    A. replacement of the control panel.
    B. replacement of the pintle O-rings.
    C. re-packing the selector body with grease.
    D. replacing the selector levers. (D2.2)

55. Technician A says that a clogged heater core could cause insufficient heater output. Technician B says that an improperly adjusted coolant control valve cable could cause insufficient heater output. Who is correct?
    A. A only
    B. B only
    C. Both A and B
    D. Neither A nor B (C1)

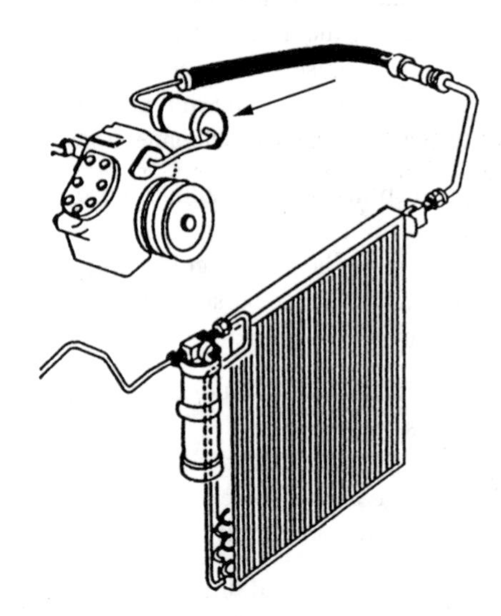

56. If the component indicated by an arrow in the figure above was not used, what would be the A/C performance result?
    A. higher A/C high-side pressure
    B. improperly filtered refrigerant
    C. evaporator icing
    D. noisier A/C operation (A4, B3.2)

57. Which of the following tests or instruments would provide the most accurate measurement of coolant antifreeze protection?
    A. hydrometer
    B. SCA test strip
    C. litmus test
    D. refractometer (C3)

58. Technician A says that the HVAC systems on most medium-duty trucks with diesel engines operate totally independently from the engine control system. Technician B says that most of the HVAC systems on these vehicles provide inputs and receive outputs from the engine control unit. Who is correct?
    A. A only
    B. B only
    C. Both A and B
    D. Neither A nor B (D1.4)

59. When refilling a truck cooling system with premixed ELC (extended life coolant) which of the following is correct practice?
    A. Add coolant conditioner directly to the premix.
    B. Only add distilled water to the premix.
    C. Top up only with propylene glycol (PG) concentrate.
    D. Never add anything to the ELC premix. (C8)

60. Technician A says that anyone who purchases R-134a must maintain records for three years indicating the name and address of the supplier. Technician B says the supplier must maintain sales records of refrigerant purchases. Who is correct?
    A. A only
    B. B only
    C. Both A and B
    D. Neither A nor B (E1)

61. Technician A says that you can test the A/C compressor clutch coil with an ohmmeter. Technician B says that connecting battery power to one terminal of the coil and grounding the other terminal can test the A/C compressor clutch coil. Who is correct?
    A. A only
    B. B only
    C. Both A and B
    D. Neither A nor B (B2.4)

62. What is the source of most noncondensable gases in refrigerant?
    A. acid
    B. air
    C. moisture
    D. oil (E5)

63. Technician A says that you should replace the receiver/drier if the sight glass appears cloudy because it indicates a ruptured dessicant pack. Technician B says that you should only replace the receiver/drier if it has a leak. Who is correct?
    A. A only
    B. B only
    C. Both A and B
    D. Neither A nor B (B3.5)

64. Technician A says most fully automatic temperature control systems have some form of self-diagnostic program that will display trouble codes. Technician B says that depending on the system design, trouble codes may be displayed digitally on the control assembly or on a hand-held scan tool. Who is correct?
    A. A only
    B. B only
    C. Both A and B
    D. Neither A nor B (D3.1)

65. Which of the following is LEAST-Likely to cause windshield fogging in the DEFROST mode?
    A. a leaking heater core
    B. a clogged evaporator drain
    C. a water leak into the plenum chamber
    D. moisture in the refrigerant (C2)

66. A whistling noise coming from under the passenger side dash with the blower motor on high speed might indicate
    A. a clogged evaporator drain.
    B. a cracked evaporator case.
    C. a broken blend door cable.
    D. a low refrigerant charge. (B3.9)

67. During an HVAC performance test, the technician hears the A/C compressor clutch slip briefly upon engagement. The Most-Likely cause is
    A. a worn out compressor clutch coil.
    B. a defective A/C compressor clutch relay.
    C. the compressor clutch air gap is too large.
    D. the compressor clutch bearing is worn. (A2, A4)

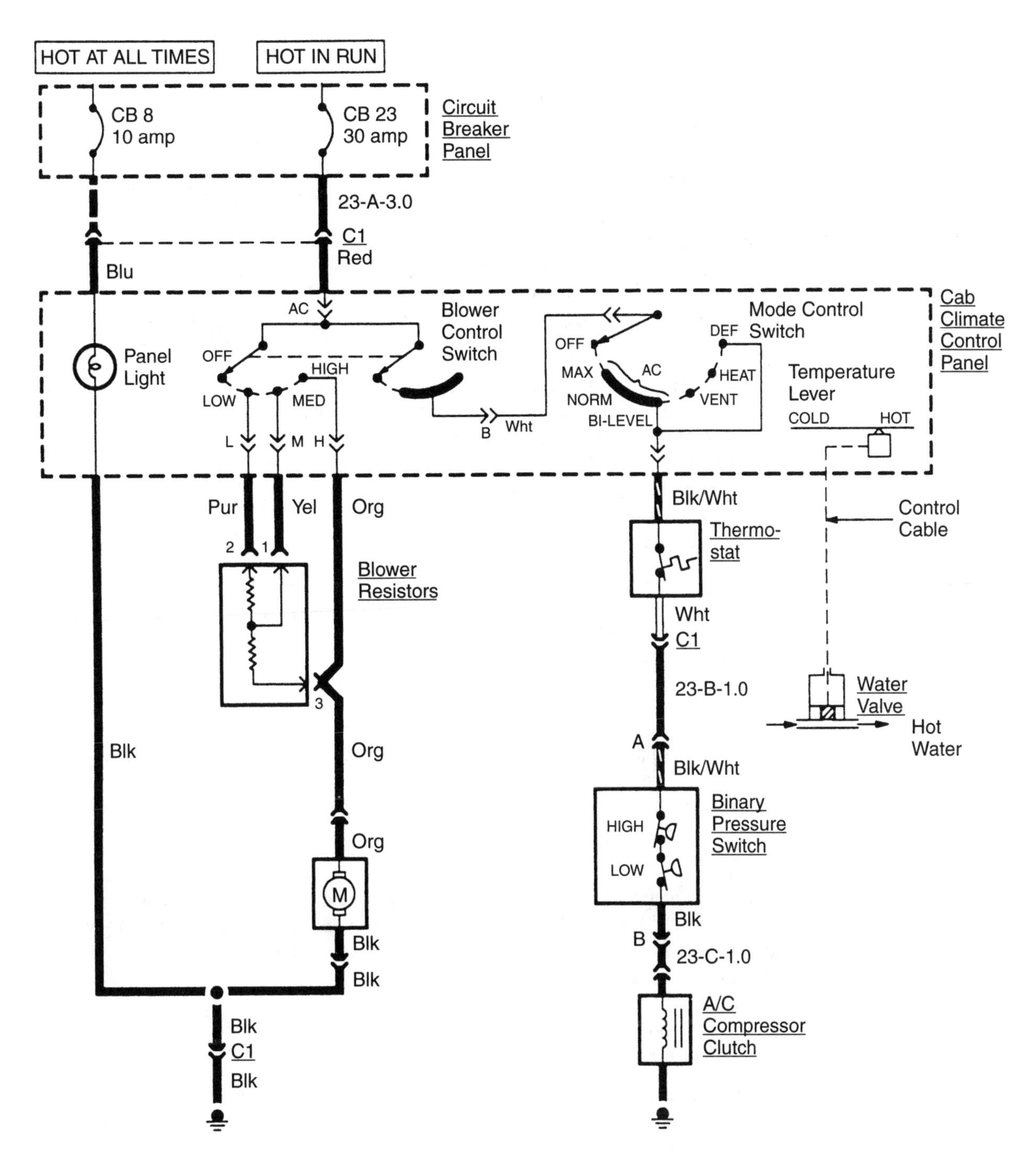

68. In the figure above, the compressor clutch is inoperative in the defrost mode but operates properly in all other A/C modes. Technician A says the low side of the binary switch may have an open circuit. Technician B says the defrost switch contacts may have an open circuit. Who is correct?

A. A only
B. B only
C. Both A and B
D. Neither A nor B

(D1.3)

69. Technician A says that an electronic blend door motor uses a pulse-width modulated (PWM) signal to control the position of the door. Technician B says that an electronic mode door motor uses a feedback device to indicate the position of the door. Who is correct?
    A. A only
    B. B only
    C. Both A and B
    D. Neither A nor B (D1.6)

70. The A/C compressor high pressure relief valve
    A. is calibrated by shimming it to the proper depth.
    B. must be replaced if it ever vents refrigerant from the system.
    C. will reset itself when A/C system pressure returns to a safe level.
    D. is not used in R-134a systems. (B3.11)

71. Technician A says you can complete the high-side charging procedure with the engine running. Technician B says if liquid refrigerant enters the compressor, damage will result to the compressor. Who is correct?
    A. A only
    B. B only
    C. Both A and B
    D. Neither A nor B (B1.8)

72. An A/C system with excessive high-side pressure could be the result of all of the following EXCEPT
    A. an overcharge of refrigerant.
    B. an overheated engine.
    C. restricted air flow through the condenser.
    D. ice buildup on the orifice tube screen. (B1.3)

73. A faint ether-like odor coming from the panel vents in the NORMAL A/C mode could indicate
    A. the evaporator core is leaking R-134a.
    B. the evaporator core is leaking R-12.
    C. the cold starting system is malfunctioning.
    D. the heater core is leaking. (A3)

74. The inside of a truck windshield has an oily film and the A/C cooling is poor. Technician A says a plugged HVAC evaporator drain may cause this oil film. Technician B says this film may be caused by a large leak in the evaporator core. Who is correct?
    A. A only
    B. B only
    C. Both A and B
    D. Neither A nor B (B3.8)

75. Oil and dirt accumulation on an A/C hose connection may indicate
    A. excessive pressure in the system.
    B. a refrigerant leak.
    C. a defective compressor shaft seal.
    D. that there is too much oil in the system. (B1.4)

76. Which method of detecting refrigerant leaks is not recommended?
    A. looking for bubbles after spraying with a soapy solution
    B. using an electronic leak detector
    C. using a flame-type leak detector
    D. using a black light detector (B1.5)

77. Each of these is an indication that the thermostat has opened EXCEPT
    A. visible coolant flow in the upper radiator tank.
    B. the upper radiator hose is hot.
    C. the lower radiator hose is hot.
    D. the temperature gauge has stabilized in the normal range. (C7)

78. What is the LEAST-Likely cause of the HVAC mode switch to function incorrectly?
    A. vacuum leak
    B. air leak
    C. broken cable
    D. open blower resistor (D2.1)

79. Technician A says that the manifold gauge set is the only tool needed to diagnose HVAC systems. Technician B says a small thermometer is a valuable tool to evaluate the performance of an HVAC system. Who is correct?
    A. A only
    B. B only
    C. Both A and B
    D. Neither A nor B (A4)

80. A screen is located in the orifice tube of an A/C system. Technician A says that the screen is a filter used to prevent particulate from circulating through the system. Technician B says that the screen is used to improve atomization of the refrigerant. Who is correct?
    A. A only
    B. B only
    C. Both A and B
    D. Neither A nor B (B3.7)

81. When the engine cooling fan viscous clutch is disengaged
    A. the fan blade will remain stationary.
    B. the engine idle will drop.
    C. the radiator shutters must be closed.
    D. the fan blade may freewheel at a reduced speed. (C9)

82. A truck technician discovers a heater core leak in the sleeper that requires its replacement. Technician A says that it is important to add the proper amount of refrigeration oil before installation. Technician B says that a PAG-based lubricant is used in modern heater cores. Who is correct?
    A. A only
    B. B only
    C. Both A and B
    D. Neither A nor B (C11)

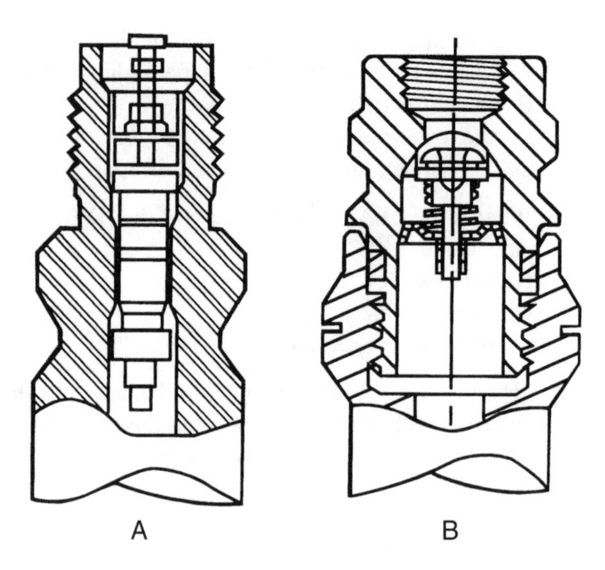

83. Technician A says that a system with service port fitting like figure A above is designed to use R-134a as a refrigerant. Technician B says that the service port fitting represented by figure B is for use with R-12. Who is correct?
    A. A only
    B. B only
    C. Both A and B
    D. Neither A nor B (A4, B3.10)

84. Which of the following would be the most common maintenance problem in truck heating systems?
    A. clogged heater cores
    B. defective control units
    C. faulty control valves
    D. coolant leaks (C3)

85. A signal to the ECM from which of the following sensors could cause the A/C compressor to disengage?
    A. engine coolant temperature (ECT) sensor
    B. intake air temperature (IAT) sensor
    C. heated oxygen sensor (HO_2S)
    D. cooling fan control sensor (D1.4)

86. A vacuum operated blend door actuator diaphragm
    A. is designed to bleed vacuum at a rate of 3 in. Hg per hour.
    B. contains a small electric motor to return the actuator to the normal position.
    C. should hold vacuum for at least one minute.
    D. is porous to allow moisture to evaporate. (D2.4)

87. Technician A says that the compressor oil needs to be checked when there is evidence of loss of system oil. Technician B says that when replacing refrigerant oil, it is important to use the specific type and quantity of oil recommended by the compressor manufacturer. Who is correct?
    A. A only
    B. B only
    C. Both A and B
    D. Neither A nor B (B1.9)

88. An ATC control panel is suspected of causing an A/C performance problem in a tractor. Technician A says to replace the ATC control computer. Technician B says to perform the sequential diagnostic steps in the truck manufacturer's service manual before replacing components. Who is correct?
    A. A only
    B. B only
    C. Both A and B
    D. Neither A nor B (D3.6)

89. All of the following statements about the chassis air system are true EXCEPT
    A. it is important to keep water drained from the system.
    B. system air can be used to operate the fan clutch.
    C. system air can be used to operate the radiator shutters.
    D. system air actuates the engine thermostat. (C9)

90. During an HVAC performance test, the technician notices that the A/C compressor outlet is nearly as hot as the upper radiator hose. Technician A says that this is a normal condition. Technician B says that the system is overcharged. Who is correct?
    A. A only
    B. B only
    C. Both A and B
    D. Neither A nor B (A3)

91. Technician A says that the best way to ensure that an A/C compressor has the proper amount of lubricant is to drain the compressor and add lubricant to the manufacturer's specifications. Technician B says if you have a doubt about the lubricant level, add an oil charge to the system. Who is correct?
    A. A only
    B. B only
    C. Both A and B
    D. Neither A nor B (B2.5)

92. A check valve is installed in-line to the vacuum reservoir to
    A. delay vacuum to downstream components.
    B. switch vacuum on and off to various components.
    C. prevent a vacuum drop during periods of low-source vacuum.
    D. monitor engine vacuum. (D2.5)

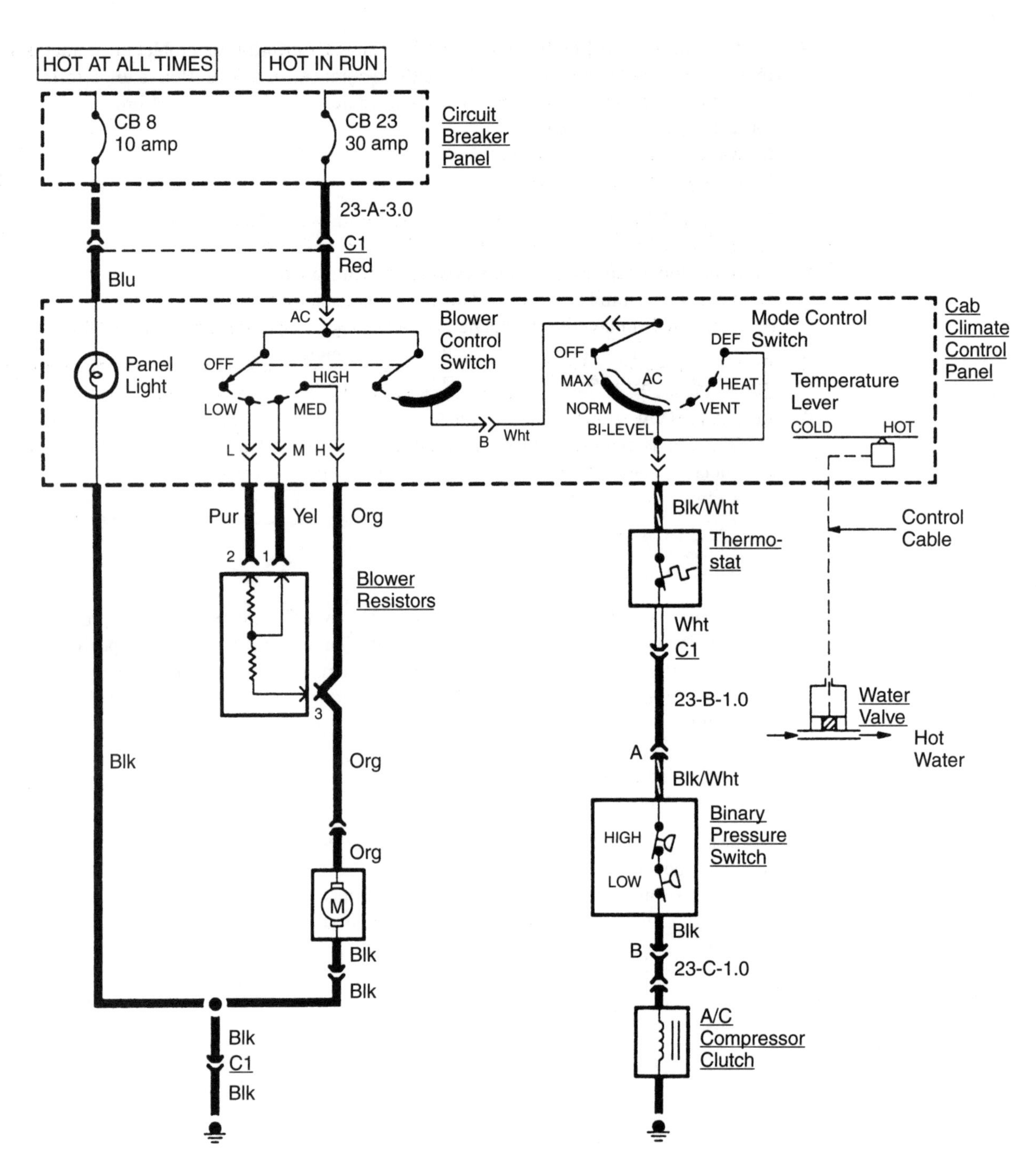

93. Refer to the figure above. All of the following could prevent A/C compressor clutch engagement EXCEPT
   A. a defective blower switch.
   B. an open binary pressure switch.
   C. a poor connection at the A/C thermostat.
   D. an open circuit breaker at CB 8. (B2.4)

94. In an ATC system, a grinding noise comes from under the dash when the temperature setting is changed from WARM to COLD. The Most-Likely cause of this problem is
    A. grit in the condenser housing.
    B. a faulty mode door actuator.
    C. bad drive gears in the blend door motor.
    D. arcing in the control head. (D3.5)

95. To prevent overfilling of recovery cylinders, the service technician must
    A. monitor cylinder pressure as the cylinder is being filled.
    B. monitor cylinder weight as the cylinder is being filled.
    C. make sure cylinder safety relief valves are in place and operational.
    D. occasionally shake the cylinder and observe any change of pressure while filling. (E2)

96. Mobile A/C systems using R-12 have flared and threaded service ports, with the port size differentiating high side from low side. Service fittings on systems with R-134a use
    A. compression fittings.
    B. SAE-approved quick-connect couplings.
    C. the same fittings as R-12 systems.
    D. 10 mm threaded ports. (A4)

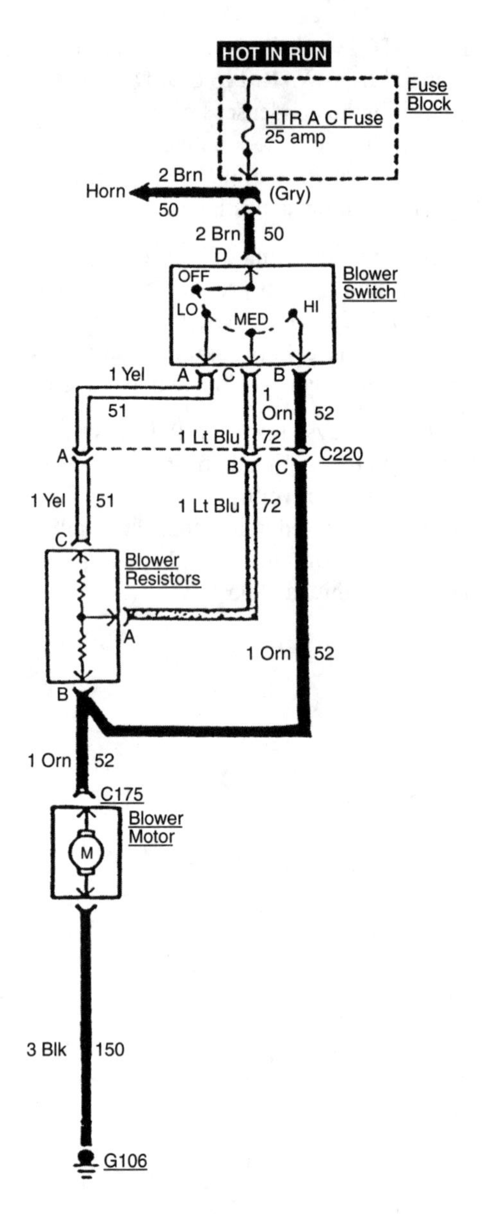

97. Refer to the figure above. The blower motor works in LO and HIGH positions but does not work on MEDIUM. Technician A says the problem could be an open blower resistor. Technician B says the problem could be the blower motor switch. Who is correct?
    A. A only
    B. B only
    C. Both A and B
    D. Neither A nor B (D1.2)

98. Technician A says that you should test the coolant condition before replacing the coolant conditioner cartridge. Technician B says that the coolant conditioner cartridge should be replaced every time the engine lube and fuel filters are changed. Who is correct?
    A. A only
    B. B only
    C. Both A and B
    D. Neither A nor B (C3)

99. While checking an ATC system with a handheld digital diagnostic reader (DDR), the ATC module generates a fault code with a failure mode identifier (FMI) of 12 described as a "defective intelligent device." The technician should
    A. look for an open wire.
    B. look for a short circuit to ground.
    C. test the control module.
    D. look for a loose connector. (D3.7)

100. Technician A says that the A/C compressor high pressure relief valve must be replaced if it operates due to extreme pressure in the A/C system. Technician B says that the compressor must be replaced if the high pressure relief valve operates. Who is correct?
    A. A only
    B. B only
    C. Both A and B
    D. Neither A nor B (B3.11)

101. An HVAC system with a vacuum control panel produces no cold air out of the dash nozzles, only out of the defrost outlets. Which of these items is the Most-Likely cause?
    A. a leaking dash vacuum switch
    B. a defective A/C compressor
    C. loss of vacuum to the control panel
    D. a defective heater control valve (D2.2)

102. A binary pressure switch provides which of the following
    A. only low pressure protection for the A/C system
    B. only high pressure protection for the A/C system
    C. both low- and high pressure protection for the A/C system
    D. neither low pressure nor high pressure protection for the A/C system (B2.2)

103. Technician A says a system having high low-side pressure accompanied by a continuously running compressor indicates that the expansion valve is stuck open. Technician B says a system having a high low-side reading accompanied by a continuously running compressor indicates a problem with the compressor clutch. Who is correct?
    A. A only
    B. B only
    C. Both A and B
    D. Neither A nor B (B1.3)

104. What is the function of the orifice tube?
    A. It allows water to drain from the evaporator case.
    B. It allows refrigerant to be metered into the evaporator.
    C. It regulates refrigerant flow through the condenser.
    D. It regulates air flow through the evaporator. (B3.7)

105. A cooling system pressure tester can be used to test
    A. thermostats.
    B. radiators, pressure caps, and hoses.
    C. A/C leaks.
    D. the blend door actuator diaphragm. (C5)

106. When evacuating an A/C system, the vacuum pump should be operated a minimum of
    A. twenty minutes.
    B. ten minutes.
    C. fifteen minutes.
    D. thirty minutes. (B1.6)

107. Technician A says that the best way to test a vacuum-operated coolant control valve is to disconnect the vacuum hose and observe whether coolant flows through it. Technician B says a vacuum-operated coolant control valve can be tested by applying and releasing vacuum and checking if the valve arm moves freely both ways. Who is correct?
    A. A only
    B. B only
    C. Both A and B
    D. Neither A nor B (C10)

108. Which of these A/C gauge set readings would Most-Likely indicate that the A/C compressor reed valves are worn out?
    A. high-side pressure too low and low-side pressure too high
    B. no pressure on either gauge
    C. both gauges read pressure that is too low
    D. both gauge readings are too high (B2.6)

109. Which of these is the Most-Likely problem when you are unable to change heater air flow from the defroster ducts to the floor outlets?
    A. The water control valve is stuck closed.
    B. There may be small objects blocking the mode door.
    C. There may be an inoperative blower motor.
    D. The blend door may be stuck. (D2.6)

110. Technician A says that moisture in an air conditioning system can combine with the refrigerant to form harmful acids. Technician B says that moisture may be removed from an air conditioning system by evacuating. Who is correct?
    A. A only
    B. B only
    C. Both A and B
    D. Neither A nor B (B1.6)

111. The A/C high pressure switch is used to
    A. boost the system high-side pressure.
    B. open the circuit to the A/C compressor clutch coil when the high-side pressure reaches its upper limit.
    C. ensure that system pressure remains at the upper limit.
    D. vent refrigerant from the compressor in the event of extremely high system pressure. (B2.2)

112. Filters installed in the HVAC air delivery system are
    A. always made of fiberglass mesh to resist corrosion.
    B. designed to remove moisture from cab and sleeper air.
    C. not individually replaceable.
    D. designed to remove dust and dirt from cab and sleeper air. (B3.9)

113. While conducting a performance test on a semiautomatic HVAC system, a Technician finds that only a small amount of air is directed to the windshield in defrost mode. The Most-Likely cause of this problem is
    A. a defective microprocessor.
    B. a defective blend door actuator.
    C. an open blower motor resistor.
    D. an improperly adjusted mode door cable. (A4)

114. In an ATC A/C system, the temperature control is set at 70°F (21°C), and the in-cab temperature is 80°F (27°C) after driving one hour. All refrigerant pressures are normal. Technician A says the in-cab sensor may be defective. Technician B says the temperature blend door may be sticking. Who is correct?
   A. A only
   B. B only
   C. Both A and B
   D. Neither A nor B (D3.1)

115. Technician A says that in a typical truck A/C system, refrigerant enters the compressor as a low pressure gas. Technician B says that in an A/C circuit, refrigerant is designed to boil in the condenser heat exchanger. Who is correct?
   A. A only
   B. B only
   C. Both A and B
   D. Neither A nor B (A4)

116. A customer complains that the air operated heater outputs hot air when the temperature control lever is in the COLD position. The LEAST-Likely cause of this problem is
   A. a defective coolant control valve.
   B. a defective engine coolant temperature sensor.
   C. a defective air control solenoid.
   D. a defective blend air door control cylinder. (D2.1)

117. A gasoline engine in a school bus with indirect port fuel injection and air conditioning has a rich air–fuel ratio. This problem could be caused by which of these items?
   A. engine overheating
   B. a defective radiator cap
   C. the engine thermostat stuck open
   D. the coolant control valve stuck open (C7)

118. If the expansion valve is not opening, the system will show which of the following pressure combinations on a manifold gauge set?
   A. low low-side pressure and low high-side pressure
   B. low low-side pressure and high high-side pressure
   C. high low-side pressure and high high-side pressure
   D. high low-side pressure and low high-side pressure (B1.3)

119. Technician A says that according to new environmental laws, shut-off valves must be placed in the closed position every time the A/C system is turned off. Technician B says that according to new environmental laws, shut-off valves must be located no more than 12 inches from test hose service ends. Who is correct?
   A. A only
   B. B only
   C. Both A and B
   D. Neither A nor B (E1)

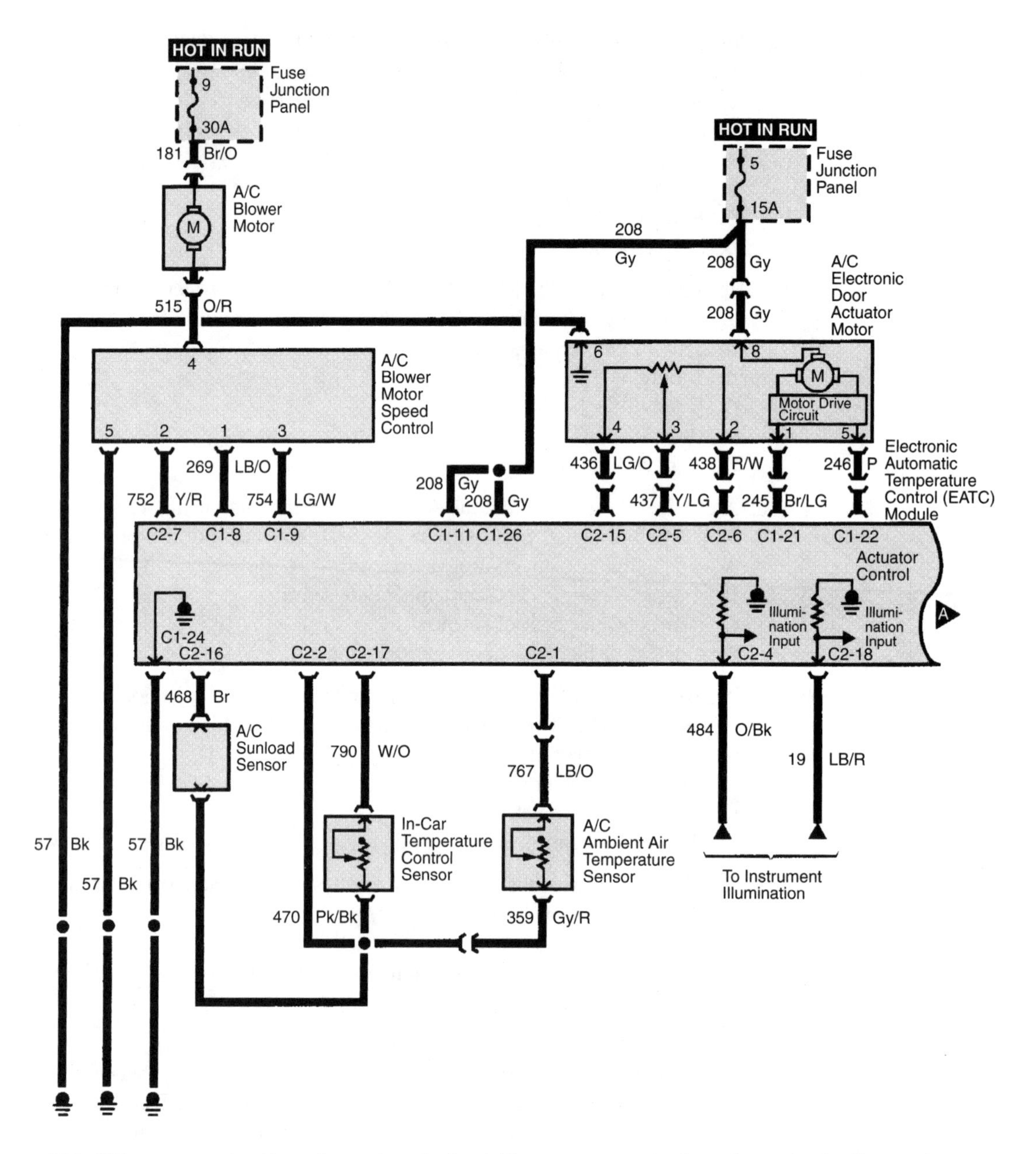

120. When measuring the voltage drop in the A/C computer ground, as shown in the figure above, connect a voltmeter from computer terminal C1-24 internal ground to the external ground. With the ignition on, the maximum allowable voltage drop should be which of the following?
   A. 0.1 volts
   B. 0.3 volts
   C. 0.5 volts
   D. 0.8 volts (D3.1)

121. On a cycling clutch A/C system, the low-side reading is too high and the high-side reading is too low. Technician A says that the compressor may have a faulty reed valve. Technician B says that an overcharge of refrigerant oil is a possible cause. Who is correct?
   A. A only
   B. B only
   C. Both A and B
   D. Neither A nor B (B1.3)

122. The test for noncondensable gases in recovered/recycled refrigerant involves
   A. comparing the pressure of the recovered refrigerant in the container to the theoretical pressure of pure refrigerant at a given temperature.
   B. comparing the atmospheric pressure to the relative humidity.
   C. comparing the container pressure with the size of the container.
   D. testing the refrigerant with a halogen leak detector. (E5)

123. A faint hissing noise is heard from the area of the evaporator immediately after shutting down the engine with the compressor clutch engaged. Technician A says that the A/C system has a leak. Technician B says the thermal expansion valve (TXV) is defective. Who is correct?
   A. A only
   B. B only
   C. Both A and B
   D. Neither A nor B (A2)

124. Before disposing of an empty or near empty original container that was used to ship refrigerant from the factory you should perform which of the following?
   A. Clean it and keep it for storage of recycled refrigerant.
   B. Open the valve completely and paint an X on the cylinder.
   C. Flush it with oil and nitrogen to keep it from rusting.
   D. Recover remaining refrigerant, evacuate cylinder, and mark it empty. (E4)

125. The driver of a gasoline powered truck notices that when the engine is turned off or the truck goes up hill, the air coming out of the A/C dash ducts immediately shifts to the defrost mode. Which of these could be more likely the cause?
   A. a defective reservoir check valve
   B. a defective vacuum pump
   C. defective fuel injectors
   D. leaking actuator diaphragm (D2.6)

126. All of these facts about a computer controlled A/C system are correct EXCEPT
   A. some actuator motors are calibrated automatically in the self-diagnostic mode.
   B. A/C diagnostic trouble codes (DTCs) represent the exact fault in a specific component.
   C. the actuator control rods must be calibrated manually on some systems.
   D. the actuator motor control rods should only require adjustment after motor replacement or adjustment. (D1.6)

127. When diagnosing a computer controlled ATC A/C system, a diagnostic trouble code is displayed indicating a fault in the temperature-blend door-actuator motor. Technician A says the first step in the repair process is to replace the temperature-blend door actuator. Technician B says you need to check the temperature blend door for binding. Who is correct?
   A. A only
   B. B only
   C. Both A and B
   D. Neither A nor B (D3.5)

128. All of the following are a type of mobile HVAC system EXCEPT
   A. blend air type.
   B. automatic temperature control (ATC).
   C. recirculating chilled and heated water.
   D. semiautomatic temperature control (SATC). (A4)

129. Technician A says that the best way to repair wiring in the compressor clutch circuit is by soldering. Technician B says that twisting the wires together and securing them with electrical tape is sufficient. Who is correct?
   A. A only
   B. B only
   C. Both A and B
   D. Neither A nor B (D1.3)

130. All of these statements about cooling system service are true EXCEPT
    A. when the cooling system pressure is increased the boiling point increases.
    B. the boiling point decreases when more antifreeze is added to the coolant.
    C. good quality ethylene glycol antifreeze contains antirust inhibitors.
    D. coolant solutions should be recovered, recycled, and handled as hazardous waste. (C8)

131. With the selector lever in the MAX A/C position, a blend air HVAC system outputs cold air for about 15 minutes, after which the output air becomes warm. The most likely cause of this problem is
    A. a defective thermal expansion valve.
    B. a defective coolant control valve.
    C. a defective blend air door return spring.
    D. a defective fresh air door. (A4)

132. Technician A says you check the calibration accuracy of an ATC system using A/C pressure gauges only. Technician B says you can recalibrate all ATC systems in the field. Who is correct?
    A. A only
    B. B only
    C. Both A and B
    D. Neither A nor B (D3.1)

133. The easiest way to check the calibration of an ATC system is to
    A. use an ohmmeter to measure the resistance of the ATC sensor.
    B. measure the voltage drop across the ATC sensor.
    C. turn the calibration screw on the ATC control unit.
    D. use a thermometer to measure the cab temperature and compare the actual temperature to the set temperature. (D3.6)

134. Technician A says you test control panel vacuum systems by applying vacuum with a hand pump to the upstream (output) end of the system. Technician B says when you supply vacuum to an actuator with a hand pump, a satisfactory actuator will maintain a steady vacuum of 15 to 20 in. Hg. for at least one minute. Who is correct?
    A. A only
    B. B only
    C. Both A and B
    D. Neither A nor B (D2.4)

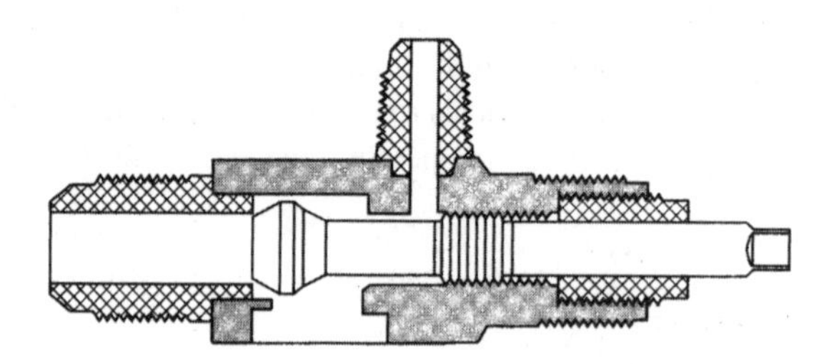

135. What is the position of the service valve in the figure above?
    A. front-seated position
    B. back-seated position
    C. mid-position
    D. normal operating position (B3.10)

136. The thermal bulb on an expansion valve must be installed in contact with
    A. the evaporator outlet tube.
    B. the condenser fins.
    C. the suction hose.
    D. the refrigerant. (B3.6)

137. A heavy-duty truck is about to have the A/C system recharged. Which of these statements is correct concerning that process?
    A. The charging process is complete when the system reaches the correct evaporator temperature.
    B. When the low side no longer moves from a vacuum to a pressure, the process is complete.
    C. The truck engine must be running during recharging.
    D. Either a high-side or a low-side charging process can be used. (B1.8)

138. Technician A says that moisture can be removed from the A/C system after charging the system with new refrigerant. Technician B says moisture that enters the A/C circuit will be harmful to the system and cause poor performance. Who is correct?
    A. A only
    B. B only
    C. Both A and B
    D. Neither A nor B (B1.6)

139. All of the statements about replacing an HVAC control panel are true EXCEPT
    A. you remove the negative battery cable before control panel service.
    B. you must recover the refrigerant before removing the control panel.
    C. if the truck contains a Supplemental Restraint System, wait the specified period after you remove the negative battery cable.
    D. self-diagnostic tests may indicate a defective control panel in an ATC system. (D1.7)

140. Before replacing an HVAC electronic control panel, the technician should
    A. remove the control cables from the vehicle.
    B. disconnect the batteries.
    C. disassemble the dash panel.
    D. apply dielectric grease to the switch contacts. (D1.7)

141. The LEAST-Likely cause of the high pressure relief valve tripping open is
    A. improper radiator shutter operation.
    B. a clogged condenser.
    C. inoperative cooling fan clutch.
    D. a defective A/C compressor. (B2.1)

142. All of the following statements about nitrogen flushing the A/C system are true EXCEPT
    A. the technician should install a pressure regulator on the supply tank.
    B. the technician should disconnect the A/C compressor.
    C. the technician should remove restrictive components (i.e., STV, TXV valve) from the system.
    D. nitrogen must not be allowed to escape into the atmosphere. (B1.7)

143. Technician A says that many trucks are programmed with engine protection strategy to warn the driver if coolant level and temperature are not within preset parameters. Technician B says that imminent engine shutdown can be overridden for a brief period after the alarm to allow the driver to move the vehicle to a safe place. Who is correct?
    A. A only
    B. B only
    C. Both A and B
    D. Neither A nor B (D1.5)

144. Which of the following modes will provide least effective windshield defrosting in the DEFROST mode?
    A. 50 percent of the output air is directed to the floor and 50 percent to the windshield
    B. 15 percent of the output air is directed to the floor and 85 percent to the windshield
    C. 25 percent of the output air is directed to the floor and 75 percent to the windshield
    D. 30 percent of the output air is directed to the floor and 70 percent to the windshield (A4)

145. The refrigerant line leading from the evaporator to the compressor contains refrigerant as a
   A. low pressure gas.
   B. low pressure liquid.
   C. high pressure gas.
   D. high pressure liquid. (A.4, B3.8)

146. The A/C compressor clutch will not engage in any mode. The clutch engages when a technician installs a jumper wire across the terminals of the low pressure cut-out switch connector. Technician A says that the low pressure cut-out switch must be defective. Technician B says that the refrigerant charge might be low. Who is correct?
   A. A only
   B. B only
   C. Both A and B
   D. Neither A nor B (B2.1)

147. The A/C condenser has several bent fins and a moderate accumulation of dead insects. Technician A says that the condenser should be replaced or it will cause the high pressure relief valve to trip. Technician B says that the condenser fins should be straightened and the dead insects removed to optimize A/C system performance. Who is correct?
   A. A only
   B. B only
   C. Both A and B
   D. Neither A nor B (B3.3)

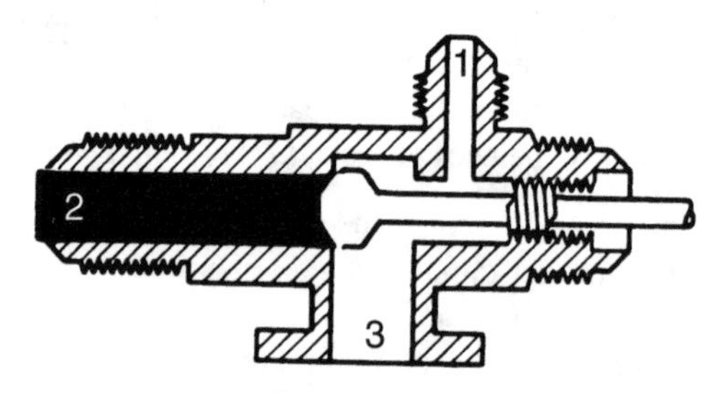

Front-Seated Position

148. With a stem type service valve in the position in the figure above
   A. you can diagnose the refrigerant system with a manifold gauge set.
   B. the refrigerant system operates normally with no pressure at the gauge ports.
   C. the refrigerant system is isolated from the compressor for compressor removal.
   D. the refrigerant system may be discharged, evacuated, and recharged. (B3.10)

149. Technician A says some manufacturers recommend the installation of an in-line filter between the evaporator and the compressor as an alternative to refrigerant system flushing. Technician B says an in-line filter containing a fixed orifice may be installed and the original orifice tube left in the system. Who is correct?
   A. A only
   B. B only
   C. Both A and B
   D. Neither A nor B (B1.7)

150. Which of the following statements about coolant control valve replacement in an ATC system is true?
   A. Refrigerant must be recovered before the coolant control valve is replaced.
   B. Heater hoses connected to the coolant control valve may be clamped during the replacement procedure to maintain coolant in the system.
   C. Access to the coolant control valve is gained through the fresh air door.
   D. The coolant control valve is an integral part of the heater hose. (D3.4)

151. The compressor discharge valve is designed to
    A. open after vaporized refrigerant is compressed, allowing the refrigerant to move to the condenser.
    B. open before vaporized refrigerant is compressed, allowing the refrigerant to move to the evaporator.
    C. regulate system variable pressure.
    D. regulate A/C system temperature. (B2.6)

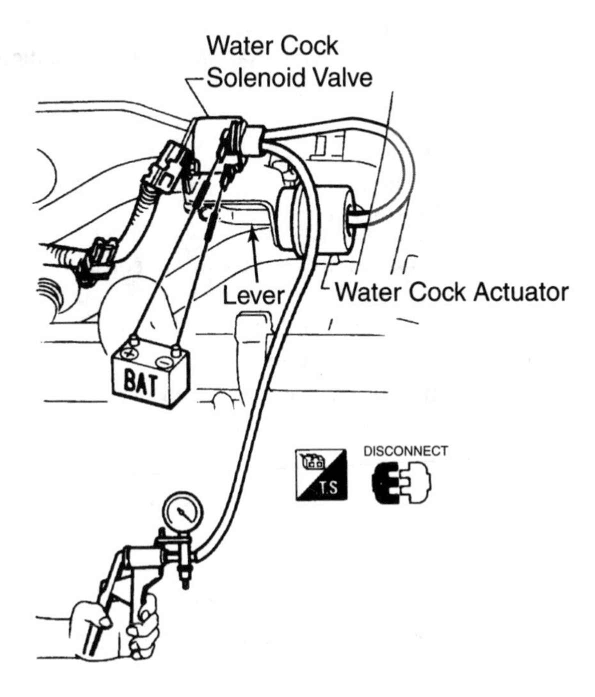

152. In the figure above vacuum is being supplied from a hand pump to the water valve solenoid and battery voltage is supplied to the solenoid terminals as shown, resulting in an audible click. The system holds at 16 in. Hg vacuum. The water valve does not move. All of these defects may be the cause of this problem EXCEPT
    A. a seized water control valve.
    B. a plugged vacuum hose between the solenoid and valve.
    C. a jammed linkage in the actuator.
    D. a seized plunger in the water valve control solenoid. (D3.4)

153. When an expansion valve is properly installed, where should the thermal bulb and capillary tube be positioned?
    A. fastened to the condenser fins using epoxy
    B. located in the accumulator
    C. held in contact with the evaporator outlet using insulating tape
    D. as an integral part of the orifice tube assembly (B3.6)

154. Which of the following statements is true about checking the A/C compressor lubricant level?
    A. The A/C system must first be evacuated for at least 30 minutes.
    B. The compressor must not be operated for at least 24 hours before checking the lubricant level.
    C. The oil level should be up to the full mark on the dipstick.
    D. The old lubricant must be measured before the new lubricant is added. (B2.5)

155. When checking the coolant condition with an SCA test strip, the technician finds that the coolant is severely over-conditioned. What should the technician do?
   A. Add more antifreeze to increase the SCA.
   B. Continue to run the truck until the next PMI.
   C. Drain the entire coolant system and add the proper SCA mixture.
   D. Run the truck with no SCA additives until the next PMI. (C3)

156. Technician A says that when evacuating a system, the vacuum pump must run at least 30 minutes to remove moisture from the system. Technician B says that observing the action of the gauges while the system is being evacuated may help indicate the presence of either a blockage or a leak. Who is correct?
   A. A only
   B. B only
   C. Both A and B
   D. Neither A nor B (B1.6)

157. In an ATC system, the blend air door constantly moves back and forth. The Most-Likely cause of this problem is
   A. a binding blend air door.
   B. a defective actuator motor.
   C. an improperly adjusted ATC sensor cable.
   D. a defective feedback device. (D3.3)

7 Appendices

Answers to the Test Questions for the Sample Test Section 5

1. D
2. A
3. D
4. C
5. D
6. D
7. D
8. C
9. C
10. D
11. C
12. A
13. A
14. B
15. C
16. C
17. A
18. C
19. D
20. B
21. C
22. D
23. C
24. C
25. D
26. B
27. D
28. A
29. A
30. B
31. A
32. C
33. A
34. C
35. A
36. B
37. B
38. C
39. A
40. A
41. B
42. A
43. B
44. D
45. A
46. C
47. C
48. B
49. A
50. B
51. B
52. D
53. B
54. D
55. A
56. C
57. D
58. A
59. C
60. C
61. C
62. C
63. C
64. A
65. B
66. D
67. A
68. C
69. C

Explanations to the Answers for the Sample Test Section 5

Question #1
Answer A is wrong. Petroleum jelly must not be used to lubricate O-rings.
Answer B is wrong. Transmission fluid can damage A/C O-rings.
Answer C is wrong. Silicone grease must not be introduced into the system.
Answer D is correct. Mineral-based refrigeration oil is the recommended lubricant for A/C O-rings.

Question #2
Answer A is correct. Only Technician A is correct. R-12 A/C service valves have a replaceable Schrader valve.
Answer B is wrong. R-134a service valves cannot be rear seated.
Answer C is wrong. Only Technician A is correct.
Answer D is wrong. Only Technician A is correct.

Question #3
Answer A is wrong. Oil must be added to the evaporator prior to installation.
Answer B is wrong. Oil must be added to the condenser prior to installation.
Answer C is wrong. Oil must be added to the accumulator prior to installation.
Answer D is correct. The addition of refrigerant oil is not required when replacing the TXV.

Question #4
Answer A is wrong. A litmus strip is used to measure relative acidity/alkalinity.
Answer B is wrong. A manometer is used to measure inlet restriction.
Answer C is correct. Antifreeze protection should be measured with a refractometer. A hydrometer can also be used to check the antifreeze protection.
Answer D is wrong. An opacity meter is used to measure smoke density.

Question #5
Answer A is wrong. Chassis air pressure can be used to operate a coolant control valve.
Answer B is wrong. Vacuum can be used to operate a coolant control valve.
Answer C is wrong. Older systems used a cable to operate the coolant control valve.
Answer D is correct. A magneto is used to generate the ignition spark in many small gasoline engines.

Question #6
Answer A is wrong. The blend door actuator does not mount using an epoxy.
Answer B is wrong. The blend door does not mount using an adhesive.
Answer C is wrong. The location varies with the truck.
Answer D is correct The blend door actuator is generally mounted to the evaporator case. The actuator connects to the blend door and controls the routing of the air through or past the heater core.

Question #7
Answer A is wrong. Electric cooling fans are often controlled by an electronic relay.
Answer B is wrong. A cooling fan module often controls electric cooling fans.
Answer C is wrong. A multiplexed electronic module often controls electric cooling fans.
Answer D is correct. Electric cooling fan motors are never controlled by an air solenoid controller.

Question #8
Answer A is wrong. A bent mounting bracket can cause belt misalignment.
Answer B is wrong. A faulty clutch bearing can cause belt misalignment.
Answer C is correct. Shimming the clutch plate sets the air gap, and air gap will not affect belt alignment.
Answer D is wrong. A faulty idler pulley will cause belt misalignment.

Question #9
Answer A is wrong. 20 in. Hg is not sufficient to ensure that all moisture has been removed from the tank.
Answer B is wrong. 22 in. Hg is not sufficient to ensure that all moisture has been removed from the tank.
Answer C is correct. A 27 in. Hg of vacuum is required to remove all of the moisture from the container.
Answer D is wrong. 12 in. Hg is not sufficient to ensure that all moisture has been removed from the tank.

Question #10
Answer A is wrong. You cannot adjust an expansion valve with any process.
Answer B is wrong. You cannot adjust an expansion valve with any process.
Answer C is wrong. You cannot adjust an expansion valve with any process.
Answer D is correct. The expansion valve is not an adjustable device.

Question #11
Answer A is wrong. Technician B is also correct.
Answer B is wrong. Technician A is also correct.
Answer C is correct. Both technicians are correct. The containers are color-coded and the fittings are different to avoid mixing refrigerants.
Answer D is wrong. Both technicians are correct.

Question #12
Answer A is correct. Only Technician A is correct. Some HVAC control panels contain independent switch modules such as the blower switch and the mode switch.
Answer B is wrong. Some truck control panel components are serviceable.
Answer C is wrong. Only Technician A is correct.
Answer D is wrong. Only Technician A is correct.

Question #13
Answer A is correct. PAG oil is specifically formulated to be compatible with R-134a. PAG oil is a synthetic-based lubricant that is light enough to be circulated by R-134a systems.
Answer B is wrong. R-12 systems use mineral-based lubricant.
Answer C is wrong. DEXRON is not compatible with R-134a and will damage A/C system components.
Answer D is wrong. SAE 30 will damage A/C system components.

Question #14
Answer A is wrong. The heater control valves are not operated by engine speed.
Answer B is correct. Only Technician B is correct. The heater control valves may be cable operated. Other methods to operate heater control valves are vacuum, air pressure, and electrical solenoids.
Answer C is wrong. Only Technician B is correct.
Answer D is wrong. Only Technician B is correct.

Question #15
Answer A is wrong. Restoration is a term generally applied to old buildings.
Answer B is wrong. Recovery is the process by which refrigerant is removed from a system.
Answer C is correct. The recycling process reduces contaminants used in refrigerant by using oil separation and filter core dryers.
Answer D is wrong. Reclamation is an industrial process by which refrigerant is restored to its original condition by an outside source.

Question #16
Answer A is wrong. Technician B is also correct.
Answer B is wrong. Technician A is also correct.
Answer C is correct. Both technicians are correct. A low heat problem could be caused by a misadjusted temperature control cable or by a clogged heater core.
Answer D is wrong. Both technicians are correct.

Question #17
Answer A is correct. Only Technician A is correct. Any leak from the system must be repaired as quickly as possible.
Answer B is wrong. If oil is leaking from the system, refrigerant must also be leaking from the system.
Answer C is wrong. Only Technician A is correct.
Answer D is wrong. Only Technician A is correct.

Question #18
Answer A is wrong. The low pressure cut-out switch is located in the low side of the system.
Answer B is wrong. An aneroid or BAP sensor senses atmospheric pressure and is part of the fuel delivery system, not the A/C system.
Answer C is correct.The low pressure cut-out switch (also called the pressure cycling switch) is mounted in the low side of the A/C system and opens when the pressure drops below 20–25 psi.
Answer D is wrong. The low pressure switch is typically not located in the cab or sleeper.

Question #19
Answer A is wrong. When the thermostat opens, the upper radiator hose rapidly gets warm.
Answer B is wrong. When the thermostat opens, the temperature gauge rises until it indicates normal operating temperature.
Answer C is wrong. When the thermostat is open, there is obvious circulation motion in the upper radiator tank.
Answer D is correct. Thermostat opening is often not noticeable in the surge tank.

Question #20
Answer A is wrong. To avoid repeat repairs, the temperature control lever should not be repaired, only replaced.
Answer B is correct. Only Technician B is correct. A control head with a broken control head should be replaced because this component receives a great deal of pressure when being operated.
Answer C is wrong. Only Technician B is correct.
Answer D is wrong. Only Technician B is correct.

Question #21
Answer A is wrong. The high pressure hose is connected to the high-side service valve.
Answer B is wrong. The low pressure hose is connected to the low-side service valve.
Answer C is correct. When the system is completely empty, the technician connects the manifold gauge center hose to a vacuum pump. The center hose is typically the yellow hose in a manifold gauge set.
Answer D is wrong. Connecting the wrong hose to the vacuum pump can result in an inadequate evacuation.

Question #22
Answer A is wrong. Many HVAC systems do not use vacuum at all.
Answer B is wrong. This describes the purpose of a vacuum delay valve.
Answer C is wrong. Vacuum actuators are not damaged by sudden changes in vacuum.
Answer D is correct. A vacuum check valve prevents loss of vacuum to components during periods of low engine vacuum.

Question #23
Answer A is wrong. A faulty circuit breaker will deny power to the system.
Answer B is wrong. Ambient air temperature below 40°F will open the ambient cut-off switch, disabling the compressor clutch.
Answer C is correct. If the ambient cut-off switch is stuck closed, the compressor clutch would not be disabled.
Answer D is wrong. An open black-yellow wire will disable the compressor clutch.

Question #24
Answer A is wrong. The blower uses three different resistance branches to achieve the different speeds, but in this switch position only one is used.
Answer B is wrong. The blower uses three different resistance branches to achieve the different speeds, but in this switch position only one, not two, are used.
Answer C is correct. Only one resistor is used to achieve medium-2 position speed.
Answer D is wrong. The absence of resistance in the circuit would produce a high, not medium speed.

Question #25
Answer A is wrong. A blown fuse or other faults could cause the problem.
Answer B is wrong. Feedback from the resistors could not cause this symptom.
Answer C is wrong. Neither technician is correct.
Answer D is correct. Neither technician is correct. The switch needs to be checked for the correct input before condemning the switch. The blower resistors have no affect on the blower switch opening and closing.

Question #26
Answer A is wrong. The evaporator drain is the most likely place to detect refrigerant leaking from the evaporator core.
Answer B is correct. The block-type expansion valve is outside of the evaporator case.
Answer C is wrong. Refrigerant leaking from the evaporator core could possibly be detected at the panel vents.
Answer D is wrong. Refrigerant leaking from the evaporator core could possibly be detected at the defrost vents.

Question #27
Answer A is wrong. A cracked fan blade should be replaced, not welded. There is too much danger that the weld could break loose and damage the radiator or even injure a person near the truck.
Answer B is wrong. A technician should never attempt to repair a cracked fan blade.
Answer C is wrong. Neither technician is correct.
Answer D is correct. Neither technician is correct. Cracked fan blades are never to be welded or repaired with epoxy. Always replace a cracked blade with a new one.

Question #28
Answer A is correct. Only Technician A is correct. When the recirculation door is in position A, outside air is drawn into the HVAC case. A noticeable change in air flow should occur when the fresh/recirculate control is changed.
Answer B is wrong. In position A, in-vehicle air is blocked.
Answer C is wrong. Only Technician A is correct.
Answer D is wrong. Only Technician A is correct.

Question #29
Answer A is correct. Only Technician A is correct. The control panel should be replaced when one segment of the digital readout on an ATC control panel is inoperative. Most of the ATC control panels allow no parts to be replaced, except illumination bulbs.
Answer B is wrong. LEDs on control panels cannot be replaced in the field.
Answer C is wrong. Only Technician A is correct.
Answer D is wrong. Only Technician A is correct.

Question #30
Answer A is wrong. A bulge in the hose indicates a weak spot.
Answer B is correct. Only Technician B is correct. The hose should be replaced immediately. All of the coolant hoses should be thoroughly checked at this time. If the hoses are the same age as the failed hose, it is advisable to replace all of them at the same time.
Answer C is wrong. Only Technician B is correct.
Answer D is wrong. Only Technician B is correct.

Question #31
Answer A is correct. The CAA established the rule of certification. This rule states that all persons authorized to operate the equipment must be certified under the Act.
Answer B is wrong. This would be in violation of the CAA.
Answer C is wrong. This would be in violation of the CAA.
Answer D is wrong. This would be in violation of the CAA.

Question #32
Answer A is wrong. Technician B is also correct.
Answer B is wrong. Technician A is also correct.
Answer C is correct. Both technicians are correct. The binary pressure switch prevents compressor operation if the refrigerant charge has been lost or ambient temperature is too cold. The binary switch also protects the system from excessive pressure.
Answer D is wrong. Both technicians are correct.

Question #33
Answer A is correct. Air cylinders are used to close shutters; they are opened by spring force.
Answer B is wrong. Coolant control valves can be operated by chassis air.
Answer C is wrong. In air-controlled HVAC systems, mode and blend air cylinders typically operate air doors.
Answer D is wrong. Air leaks can cause mode and blend air doors to react sluggishly or to be inoperative.

Question #34
Answer A is wrong. Technician B is also correct.
Answer B is wrong. Technician A is also correct.
Answer C is correct. Both technicians are correct. R-12 is distributed in containers that are white and R-134a is distributed in containers that are blue. The service fittings on R-134a containers are larger than on R-12 containers to help prevent cross contamination.
Answer D is wrong. Both technicians are correct.

Question #35
Answer A is correct. Only Technician A is correct. If the air gap is too great, the clutch will slip. Compressor clutch air gap can be checked by sliding a feeler gauge between the drive and driven plates. If the air gap is above specs, typically shims can be removed to lower the gap.
Answer B is wrong. The drive and driven plates must be replaced, not resurfaced.
Answer C is wrong. Only Technician A is correct.
Answer D is wrong. Only Technician A is correct.

Question #36
Answer A is wrong. Debris trapped in the condenser fins will significantly affect air flow through the condenser.
Answer B is correct. Relative humidity will not significantly affect air flow through the condenser. However, the humidity level greatly affects the pressures in the high side of the system. High humidity levels cause the pressure in the high side of the system to elevate dramatically.
Answer C is wrong. Bent fins will affect air flow through the condenser.
Answer D is wrong. Vehicle speed will affect air flow through the condenser.

Question #37
Answer A is wrong. Refrigerants are heavier than air and will not be effectively detected above the suspected leak.
Answer B is correct. Refrigerants are heavier than air and will be most effectively detected below a suspected leak.
Answer C is wrong. Six inches upstream from the leak would cause small leaks to go undetected.
Answer D is wrong. Six inches downstream from the leak would cause small leaks to go undetected.

Question #38
Answer A is wrong. Technician B is also correct.
Answer B is a wrong. Technician A is also correct.
Answer C is correct. Both technicians are correct. Hand-held diagnostic tools are often used to calibrate and to diagnose ATC systems. However, most ATC systems have a self-diagnostic feature that allows a technician to retrieve codes and even recalibrate the actuators just by pressing the correct buttons on the control head.
Answer D is wrong. Both technicians are correct.

Question #39
Answer A is correct. A DMM is the most precise tool for electrical/electronic components or systems and because of its 10 megohm impedance, it will not harm solid-state components.
Answer B is wrong. A self-powered test lamp could damage solid-state components. The self-powered test light (also called a continuity tester) is only used to check switches and wiring for continuity.
Answer C is wrong. An analog VOM can damage solid-state components because it draws too much current.
Answer D is wrong. A 12-volt test lamp can damage solid-state components because it draws too much current.

Question #40
Answer A is correct. Only Technician A is correct. A growling noise from the water pump indicates a worn bearing. If the pump is accessible at all, a technician can use a listening device such as a stethoscope or a yardstick and contact the pump housing to pick up the vibration.
Answer B is wrong. Cavitation damage will not cause this symptom. Overheating due to the lack of water circulation will result from a damaged water pump impeller.
Answer C is wrong. Only Technician A is correct.
Answer D is wrong. Only Technician A is correct.

Question #41
Answer A is wrong. A cracked mounting plate could cause drive-belt wear by not holding the compressor in place. This would misalign the compressor and cause the belt to wear.
Answer B is correct. A cracked mounting plate will not cause internal compressor damage.
Answer C is wrong. A cracked mounting plate can cause a vibration with the compressor clutch engaged. This vibration can appear to be caused by a faulty compressor but is just a cracked mounting plate.
Answer D is wrong. A cracked mounting plate can cause belt squeal or chatter because of misaligned pullies.

Question #42
Answer A is correct. When you increase the cooling system pressure, you increase the boiling point, not decrease it.
Answer B is wrong. When more antifreeze is added to the coolant mix, the boiling point is increased.
Answer C is wrong. High quality ethylene glycol antifreeze contains a corrosion inhibitor.
Answer D is wrong. Coolant solutions must be recovered, recycled, or handled as hazardous material.

Question #43
Answer A is wrong. Applying shop air to a vacuum actuator will damage the actuator.
Answer B is correct. The correct to test a vacuum actuator is to connect a hand-held vacuum pump and supply a 15 to 20 in. Hg to the actuator. The actuator rod should move full stroke and should not bleed down for one minute.
Answer C is wrong. Testing an actuator in this manner could condemn a good component.
Answer D is wrong. The evacuation pump is NOT an appropriate tool to test actuators.

Question #44
Answer A is wrong. A clogged evaporator drain will cause fogging on the inside of the windshield, accompanied by a mildew smell.
Answer B is wrong. A leaking heater core will cause fogging on the inside of the windshield, accompanied by a sweet smell.
Answer C is wrong. An iced-up evaporator core will not cause the windshield to fog. This problem would result in poor A/C cooling.
Answer D is correct. Under these conditions, a cold windshield causes moisture to condense from the outside air.

Question #45
Answer A is correct. Only Technician A is correct. When the compressor clutch is not engaged, the pressure in the A/C system equalizes. Another possible cause for the pressures to be the same is if the manifold valves are in the open position. When checking pressures, it is important to have the manifold valves closed.
Answer B is wrong. A restriction in the expansion valve will cause low, but not identical, system pressures.
Answer C is wrong. Only Technician A is correct.
Answer D is wrong. Only Technician A is correct.

Question #46
Answer A is wrong. Any pressure bleed-down indicates that the heater core is leaking and must be replaced.
Answer B is wrong. If the heater core leaks, it should be replaced immediately to prevent coolant loss and the risk of engine damage.
Answer C is correct. When you test a heater core for leaks, you apply 10 psi of air pressure, and a pressure leakage rate of 5 psi in 3 minutes is not acceptable. The heater core would need to be replaced.
Answer D is wrong. This is an acceptable method of checking a heater core for leaks.

Question #47
Answer A is wrong. Technician B is also correct.
Answer B is wrong. Technician A is also correct.
Answer C is correct. Both technicians are correct. When a compressor is replaced, the receiver/drier should be replaced. Also, whenever the A/C system has been open to the atmosphere for an extended time, the receiver/drier should be replaced. When replacing the receiver dryer, it is important to mount it last to prevent it being exposed to the atmosphere for an extended period of time.
Answer D is wrong. Both technicians are correct.

Question #48
Answer A is wrong. Raising the pressure of the cooling system raises the boiling point of the coolant.
Answer B is correct. Raising the pressure compresses the rate of gas expansion, thereby increasing the temperature to reach the boiling point. Each 1 psi of pressure added to the cooling system results in the boiling point of the coolant being raised by 3°F.
Answer C is wrong. Raising the pressure in the cooling system does not prevent corrosion. Antifreeze contains corrosion inhibiters that help prevent corrosion.
Answer D is wrong. Raising the pressure raises the boiling point 3 degrees for each psi of pressure that is added.

Question #49
Answer A is correct. Only Technician A is correct. An NTC thermistor is commonly used as a coolant temperature sensor in truck engines. NTC stands for negative temperature coefficient. As the temperature rises, the resistance decreases.
Answer B is wrong. Truck coolant sensors are usually supplied with V-Ref (±5 VDC) and not V-Bat.
Answer C is wrong. Only Technician A is correct.
Answer D is wrong. Only Technician A is correct.

Question #50
Answer A is wrong. Damaged or misaligned condenser mounts should not affect compressor operation.
Answer B is correct. Only Technician B is correct. Deformed or improperly aligned condenser mounting insulators could damage the condenser and refrigerant lines. The insulators are made from rubber and act to cushion the vibration of the truck and engine from the condenser and lines. If the insulators are deformed or misaligned, leaks could develop in these areas.
Answer C is wrong. Only Technician B is correct.
Answer D is wrong. Only Technician B is correct.

Question #51
Answer A is wrong. R-12 is an odorless gas and would not cause a sweet odor.
Answer B is correct. Only Technician B is correct. The odor of leaking antifreeze can be drawn from the evaporator case even when operating in an A/C mode. If the heater core was causing the odor, the technician would smell a sweet aroma with the HVAC controls set at any setting. Other signs of a leaking heater core might be coolant in the floorboard, a steamed-up windshield, or coolant leaking from the HVAC housing drain tube.
Answer C is wrong. Only Technician B is correct.
Answer D is wrong. OnlyTechnician B is correct.

Question #52
Answer A is wrong. Some ATC systems use a sunlight sensor to gauge the intensity of the sunlight entering the vehicle. This photo-resistor is usually located on the dash panel and sends a varying voltage to the processor as the light level changes.
Answer B is wrong. Most ATC systems use an ambient temperature sensor to allow the controller to calculate the outside air temperature.
Answer C is wrong. An evaporator temperature sensor is used to tell the controller when to disengage the compressor. If the temperature gets too cold in the evaporator, the controller will disengage the A/C compressor.
Answer D is correct. The manifold pressure sensor is an input to the diesel electronic fuel injection control unit.

Question #53
Answer A is wrong. A technician should never just use the repair order to diagnose a vehicle. Each concern on the repair order should be verified to make sure that the driver is accurate in the concern.
Answer B is correct. Only Technician B is correct. A technician should never assume that any vehicle has the problem that is on the repair order. It is always wise to verify the complaint. This vehicle should be road tested to gather as much information as possible before beginning the diagnosis.
Answer C is wrong. Only Technician B is correct.
Answer D is wrong. Only Technician B is correct.

Question #54
Answer A is wrong. A low refrigerant charge will not affect heater temperature control.
Answer B is wrong. Even with a very low coolant level, cooling will take place with the A/C on.
Answer C is wrong. A compressor clutch failure will not affect heater temperature control.
Answer D is correct. A broken blend door cable will prevent the control of the HVAC system output air. The blend door moves to route the air through or around the heater core. When the temperature lever is set to hot, the blend door routes all of the air past the heater core. When the temperature lever is in the full cold position, the blend door blocks any air from passing by the heater core.

Question #55
Answer A is correct. Vinegar will kill the mold and mildew that is the source of the odor. Another method to correct a mildew odor is to use a fungicide spray on the evaporator core. This can be applied by removing the blower resistor and spraying the surface of the evaporator core. If this does not correct the problem, the evaporator core should be replaced.
Answer B is wrong. This method will simply mask the odor and not eliminate the problem.
Answer C is wrong. This method poses a possible fire hazard.
Answer D is wrong. This method will simply mask the odor and not eliminate the problem.

Question #56
Answer A is wrong. Technician B is also correct.
Answer B is wrong. Technician A is also correct.
Answer C is correct. Both technicians are correct. Excessive high-side pressure that caused the pressure relief valve on the A/C compressor to operate might have been the result of a faulty engine cooling fan clutch. A faulty shutter control valve could also cause the pressure relief valve to operate.
Answer D is wrong. Both technicians are correct.

Question #57
Answer A is wrong. Having a new compressor does not guarantee that a vehicle has been retrofitted.
Answer B is wrong. Although it is advisable to replace the dryer when performing a retrofit, seeing a new one on a vehicle does not mean that it has had a retrofit.
Answer C is wrong. Seeing the caps missing from the service ports does not prove that the vehicle has been retrofitted.
Answer D is correct. Whenever a vehicle is retrofitted from R-12 to R-134a, a blue retrofit decal is to be installed in visible location. The technician needs to list the charge amount, the amount and type of oil used, and the name of the service business that performed the retrofit.

Question #58
Answer A is correct. A misaligned duct could cause a whistling noise. Other possible causes of a whistling noise might be a cracked case or foreign items such as paper or leaves in the ducts or blower cage.
Answer B is wrong. A defective actuator will not cause this symptom. If an actuator was bad, the air would not come out at the correct place.
Answer C is wrong. An improperly adjusted cable will not cause this symptom. If the mode door cable was misadjusted, the air would not come out at the correct place.
Answer D is wrong. A poor connection will not cause this symptom. A poor connection at the blend door motor would result in no control of the temperature of the air.

Question #59
Answer A is wrong. A 12-volt test light can damage electronic circuit components. A test light will pull too much current to be used on electronic circuits.
Answer B is wrong. An analog multimeter can damage electronic circuit components. If a multimeter is required, it must be a DMM (digital multimeter) with high impedance.
Answer C is correct. A digital multimeter is the correct instrument to use to diagnose electronic circuit component malfunctions. One note of caution is to make sure the multimeter is a high impedance device. To safely be used on electronic circuits, the meter needs to have at least 10 megohms of internal resistance (impedance).
Answer D is wrong. An A/C charging station cannot be used to diagnose electronic circuit component malfunctions.

Question #60

Answer A is wrong. If a cylinder had 220 psi of pressure at 70 degrees, then it would definitely have some other chemical mixed with it. A technician would need to use an identifier to find out what else is mixed with the refrigerant.

Answer B is wrong. 125 psi of pressure in a cylinder of refrigerant at 70 degrees would indicate a high percentage of air. The refrigerant would need to be recovered and recycled to remove the air.

Answer C is correct. The relationship of pressure to temperature at a constant volume is direct so that at 70°F the pressure is about 70 psi. This is known as Charles' Gas Law.

Answer D is wrong. The relationship of pressure to temperature at a constant volume is direct so that at 70°F the pressure is about 70 psi, not 30 psi.

Question #61

Answer A is wrong. A technician can install refrigerant through both service valves when the engine is not running.

Answer B is wrong. A technician can install refrigerant through the low-side service valve when the engine is running.

Answer C is correct. A technician should never attempt to charge an A/C system through the high-side service valve when the engine is running because this high pressure will likely exit the vehicle, enter your charging tank, and could explode.

Answer D is wrong. A technician can charge an A/C systems directly from an approved charging station.

Question #62

Answer A is wrong. The vapor line (also called the suction line) carries low pressure vapor to the compressor inlet connection.

Answer B is wrong. The condenser removes heat from the high pressure gas, which allows it to condense back into a liquid.

Answer C is correct. The orifice tube separates the high and low side of the refrigeration system. Refrigerant enters the orifice tube as a high pressure liquid. After passing through the orifice tube, the refrigerant is at a much lower pressure so it starts to boil. This boiling allows the refrigerant to take on latent heat from the surface around it in the evaporator. The refrigerant continues to boil until it completely turns into a vapor and is routed to the compressor inlet.

Answer D is wrong. The capillary tube is a temperature-sensing device on a TXV and does not contain refrigerant.

Question #63

Answer A is wrong. The running motor indicates that the module is functioning.

Answer B is wrong. A defective feedback device will not prevent the door from moving when the motor runs. If the potentiometer was bad, the motor would run but it would not be in the proper position.

Answer C is correct. A defective drive gear in the actuator will cause this condition. This problem is usually accompanied by a clicking noise of the actuator in the dash area. The actuator must be replaced to correct this problem.

Answer D is wrong. An improperly adjusted ATC sensor cable will not cause this condition. Most ATC systems use electric actuators to move the doors, eliminating the need for cables.

Question #64

Answer A is correct. A broken blend door cable will make HVAC output temperature uncontrollable. The blend door controls the temperature of the air by routing the air through or around the heater core. When the control is set to cold, the blend door blocks the air from going past the heater core. When the control is set for hot, the blend door routes all of the air past the heater core. When the control is set anywhere else, the blend mixes the air by allowing some of the air to go past the heater core and some to bypass it.

Answer B is wrong. A defective compressor clutch will not affect heater temperature control.

Answer C is wrong. A clogged orifice tube will not affect heater temperature control.

Answer D is wrong. An inoperative blower motor has no affect on temperature control.

Question #65

Answer A is wrong. A faulty compressor discharge valve would not cause this problem. A faulty compressor discharge valve will cause poor cooling accompanied with unusual gauge pressures.

Answer B is correct. A restriction in the high pressure hose will result in a band of frost on the high pressure line. All of the components in the high side of the system should be hot while the system is operating. A restriction anywhere in the high side will result in a cold spot or even frost to appear.

Answer C is wrong. A clogged orifice tube will not cause a band of frost to appear on the high pressure hose. If the orifice tube gets restricted, then the line downstream might be frosted.

Answer D is wrong. Moisture in the system could cause a restriction at the orifice tube and result in frost appearing at the orifice tube.

Question #66

Answer A is wrong. If the compressor bearing were the cause, the noise would only be present with the clutch engaged.

Answer B is wrong. A defective clutch bearing causes a growling noise with the clutch disengaged but goes away when the clutch is turned on. The reason the noise goes away is because both parts of the bearing are being turned at the same speed when the clutch is engaged.

Answer C is wrong. Neither technician is correct.

Answer D is correct. Neither technician is correct. The growling or rumbling noise would have to be coming from something in motion with the A/C turned on or off. Items that could possibly cause this problem would include the water pump, an idler bearing, an alternator bearing, or possibly a loose mounting bracket in the drive belt system.

Question #67

Answer A is correct. A/C flush solvent, such as Dura 141, can remove debris from lines and hose assemblies that do not have mufflers in them. If the line or hose assembly has a muffler, then it cannot be flushed and must be replaced. Condensers that are used on R-134a systems have very small passages. In the event of a mechanical failure, the passages of the condenser get clogged and cannot be effectively flushed out. The condenser needs to be replaced.

Answer B is wrong. CFC refrigerants must never be used to flush the A/C system.

Answer C is wrong. In order for a filter to work, the particulate must circulate through the system, creating the potential for component damage or clogging.

Answer D is wrong. Nitrogen flushing will not clean residue from the lines and hoses. However, using pressurized nitrogen is a good method to push the flush solvent through the system.

Question #68

Answer A is wrong. This is not an acceptable method of adjusting the lubricant level in the system.

Answer B is wrong. This is not an acceptable method because the A/C system must first be recovered and the compressor removed from the vehicle.

Answer C is correct. The only accurate way to measure the amount of oil in a compressor is to remove the compressor and drain the oil and measure. If less than 2 ounces is removed from the compressor, then add 2 ounces of new oil to the compressor. If more than 2 ounces was removed, then add the amount removed of new oil. Always follow the compressor manufacturer recommendations in regards to adding oil to the compressor.

Answer D is wrong. This method will result in overfilling the compressor.

Question #69
Answer A is wrong. Technician B is also correct.
Answer B is wrong. Technician A is also correct.
Answer C is correct. Both technicians are correct. When installing a new or rebuilt compressor you should first turn it over by hand and drain any oil shipped with the compressor, and then install the correct amount of the right oil before installing it; also, some compressors are shipped without oil. Most manufacturers recommend draining the old compressor into a measuring cup. If 2 ounces or more is drained from the old compressor, then the same amount of new oil should be added to the new compressor. If less than 2 ounces is drained from the old compressor, then 2 ounces of new oil should be added to the new compressor. After adding the oil, the compressor should be turned by hand to help displace the oil to prevent damage when the compressor is first turned on.
Answer D is wrong. Both technicians are correct.

Answers to the Test Questions for the Additional Test Questions Section 6

1. C
2. A
3. A
4. D
5. D
6. D
7. B
8. A
9. B
10. C
11. A
12. B
13. A
14. B
15. A
16. A
17. D
18. D
19. C
20. B
21. C
22. A
23. D
24. C
25. C
26. B
27. C
28. A
29. A
30. A
31. C
32. B
33. A
34. A
35. C
36. B
37. A
38. A
39. D
40. C
41. A
42. C
43. C
44. D
45. C
46. B
47. C
48. B
49. A
50. B
51. D
52. C
53. B
54. A
55. C
56. D
57. D
58. B
59. D
60. B
61. C
62. B
63. A
64. C
65. D
66. B
67. C
68. B
69. B
70. C
71. B
72. D
73. A
74. B
75. B
76. C
77. C
78. D
79. B
80. A
81. D
82. D
83. D
84. D
85. A
86. C
87. C
88. B
89. D
90. A
91. A
92. C
93. D
94. C
95. B
96. B
97. B
98. A
99. C
100. D
101. C
102. C
103. A
104. B
105. B
106. D
107. B
108. A
109. B
110. C
111. B
112. D
113. D
114. C
115. A
116. B
117. C
118. A
119. B
120. A

121. A
122. A
123. D
124. D
125. A
126. B
127. B
128. C
129. A
130. B
131. A
132. D
133. D
134. B
135. A
136. A
137. D
138. B
139. B
140. B
141. D
142. D
143. C
144. A
145. A
146. B
147. B
148. C
149. D
150. B
151. A
152. D
153. C
154. D
155. C
156. C
157. D

Explanations to the Answers for the Additional Test Questions Section 6

Question #1
Answer A is wrong. A spring lock tool is not shown.
Answer B is wrong. A bearing puller is not shown.
Answer C is correct. The hose end crimping tool is shown in the figure.
Answer D is wrong. A flare tool is not shown.

Question #2
Answer A is correct. The condenser generally holds 1 ounce of refrigeration oil. Always refer to the manufacturers' recommendations before adding oil to A/C components.
Answer B is wrong. This amount is too large.
Answer C is wrong. This amount is too large.
Answer D is wrong. This amount is too large.

Question #3
Answer A is correct. The expansion valve is located at the evaporator inlet.
Answer B is wrong. This is the wrong location.
Answer C is wrong. This is the wrong location.
Answer D is wrong. This is the wrong location. The receiver/dryer is located in this location.

Question #4
Answer A is wrong. This temperature is too low.
Answer B is wrong. This temperature is too low.
Answer C is wrong. This temperature is too low.
Answer D is correct. Blower operation is delayed to prevent air from entering the cab at an uncomfortable temperature.

Question #5
Answer A is wrong. R-12 is not flammable.
Answer B is wrong. The gas that is formed is toxic, attacking the nervous system.
Answer C is wrong. R-12 does not form chlorine gas in the presence of a flame.
Answer D is correct. When R-12 is burned it creates phosgene gas. Phosgene gas is dangerous to humans.

Question #6
Answer A is wrong. Many fan clutches are air operated.
Answer B is wrong. Some fan clutches are operated by a thermostatic spring and fan clutch fluid.
Answer C is wrong. Some fans use an electrically actuated clutch.
Answer D is correct. Hydraulic switches are rarely used.

Question #7
Answer A is wrong. The back-seated position is the normal operating position.
Answer B is correct. Compressor damage would result if it were operated with the service valve in the front-seated position.
Answer C is wrong. The mid-position is used when a manifold gauge set is hooked up.
Answer D is wrong. The high-side service valve is typically hot when the compressor is running.

Question #8
Answer A is correct. The standard coolant-heated heater-core type of heater is a forced air convection system.
Answer B is wrong. An immersion heater is sometimes used to preheat engine coolant.
Answer C is wrong. Fuel-fired heaters are still uncommon in the United States.
Answer D is wrong. Electric heaters are not practical for mobile HVAC systems.

Question #9
Answer A is wrong. 5–10 psi is an extremely low low-side pressure indicative of a clogged orifice tube or low refrigerant charge.
Answer B is correct. 25–45 psi is the normal range of low-side pressures.
Answer C is wrong. 60–80 psi is a very high low-side pressure.
Answer D is wrong. 180–205 psi is a typical high-side pressure.

Question #10
Answer A is wrong. Technician B is also correct.
Answer B is wrong. Technician A is also correct.
Answer C is correct. Both technicians are correct. A sticky film on the windshield is an indication of an engine coolant leak at the heater core.
Answer D is wrong. Both technicians are correct.

Question #11
Answer A is correct. Cooling and heating system hoses do not need to be replaced because they leak at the hose clamps; usually tightening the clamps repairs the leak.
Answer B is wrong. Cracked hoses must be replaced.
Answer C is wrong. Bulging hoses must be replaced.
Answer D is wrong. Spongy hoses must be replaced.

Question #12
Answer A is wrong. Refrigerant oil must be added to the evaporator prior to installation.
Answer B is correct. Most manufacturers recommend that 3 ounces of oil be added to a new evaporator.
Answer C is wrong. Nine ounces of refrigerant oil would cause an excessive system oil level.
Answer D is wrong. The entire A/C system is likely to contain about 14½ ounces of oil.

Question #13
Answer A is correct. A saline solution will cause corrosion.
Answer B is wrong. A soft whisk broom can be used to remove debris from the condenser fins.
Answer C is wrong. Compressed air can be used to remove debris from the condenser fins.
Answer D is wrong. A soap and water solution can be used to remove debris from the condenser fins.

Question #14
Answer A is wrong. R-134a does have an ether-like odor.
Answer B is correct. R-134a has a faint ether-like odor. It is possible for customers to comment that they have an ether-like smell in their vehicle if the evaporator core has a large leak.
Answer C is wrong. R-134a does not have a strong rotten egg odor.
Answer D is wrong. R-134a does not have a cabbage-like odor.

Question #15
Answer A is correct. Restricted air flow through the condenser will cause elevated high-side pressure.
Answer B is wrong. A thermostat that is stuck open will cause poor heater performance.
Answer C is wrong. The thermal bulb will not leak.
Answer D is wrong. The bypass valve is a component in thermactor air cleaner.

Question #16
Answer A is correct. The figure does show the air mix door adjustment.
Answer B is wrong. The figure does not show the manual coolant valve adjustment.
Answer C is wrong. The figure does not show the ventilation door control rod adjustment.
Answer D is wrong. The figure does not show the defroster door control rod adjustment.

Question #17
Answer A is wrong. Only R-12 A/C systems use mineral oil.
Answer B is wrong. PAG oil is a synthetic lubricant used in R-134a A/C systems.
Answer C is wrong. Neither technician is correct.
Answer D is correct. Neither technician is correct. R-12 systems use a mineral-based lubricant and R-134a systems use PAG oil, a synthetic lubricant.

Question #18
Answer A is wrong. A kinked cable housing will cause the cable to bind.
Answer B is wrong. Corrosion in the cable housing will cause the cable to bind.
Answer C is wrong. A deformed or over-tightened cable clamp will cause the cable to bind.
Answer D is correct. The mode doors are not controlled by the temperature control cable.

Question #19
Answer A is wrong. A black light is only useful when a special dye is introduced into the cooling system.
Answer B is wrong. A leak detector is used to find refrigerant leaks.
Answer C is correct. A good visual inspection will locate most external coolant leaks.
Answer D is wrong. White smoke in the engine exhaust is evidence of burning antifreeze.

Question #20
Answer A is wrong. The sensor tip should be as close as possible to the fitting.
Answer B is correct. Refrigerant is heavier than air, so it is most easily detected just below a leaking fitting.
Answer C is wrong. The sensor probe should be just below the fitting.
Answer D is wrong. The leak could go undetected using this method. The sensor probe should be just below the fitting.

Question #21
Answer A is wrong. The engine control unit can control an electric condenser fan motor.
Answer B is wrong. The body control unit can sometimes control an electric condenser fan motor.
Answer C is correct. A manual switch is typically not used to control an electric condenser fan.
Answer D is wrong. An electric condenser fan motor can be controlled by an electronic relay.

Question #22
Answer A is correct. The engine radiator shutters are usually controlled mechanically or by the engine control module.
Answer B is wrong. The ATC system controls the speed of the blower motor.
Answer C is wrong. The ATC system controls the blend door actuator to modulate the output air temperature.
Answer D is wrong. The ATC system controls the position of the outside air door.

Question #23
Answer A is wrong. Flushing the cooling system removes rust from the system.
Answer B is wrong. Flushing the cooling system removes contaminants from the system.
Answer C is wrong. Flushing the cooling system can increase the life of the system components.
Answer D is correct. Acids of combustion are typically found in the engine oil, not in the coolant.

Question #24
Answer A is wrong. Fan clutch replacement does not affect the life of the water pump.
Answer B is wrong. The water pump generally has a longer service life than the heater hoses.
Answer C is correct. Even a small leak from the weep hole indicates that the front seal of the water pump has failed.
Answer D is wrong. Thermostat operation does not indicate a need to replace the water pump.

Question #25
Answer A is wrong. Battery removal has no effect on this service.
Answer B is wrong. This has no bearing on the removal process.
Answer C is correct. If the blend air door is binding, it could damage the new actuator.
Answer D is wrong. Blend door actuators are not like sensitive CMOS electronic components.

Question #26
Answer A is wrong. 3 psi is far too low. For every 1 psi of pressure that is added to the cooling system, the boiling point is raised 3°F.
Answer B is correct. Heavy-duty truck cooling systems are typically pressurized to about 10 psi to 15 psi.
Answer C is wrong. 20 psi is too high and could possibly cause leaks in cooling system components.
Answer D is wrong. 30 psi is very high and would likely cause leaks in cooling system components.

Question #27
Answer A is wrong. Technician B is also correct.
Answer B is wrong. Technician A is also correct.
Answer C is correct. Both technicians are correct. A malfunctioning TXV could cause the evaporator core to ice up, thus causing reduced air flow to the panel outlets.
Answer D is wrong. Both technicians are correct.

Question #28
Answer A is correct. In the MAX A/C mode the outside air door is closed. Running the A/C in the Max mode will allow the air inside the cab to re-circulate within the cab, resulting in colder duct temperatures.
Answer B is wrong. The defroster door is closed in the MAX A/C mode.
Answer C is wrong. The compressor clutch will cycle normally in the MAX A/C mode.
Answer D is wrong. The blower functions normally in the MAX A/C mode.

Question #29
Answer A is correct. A bad blower switch would be the most likely cause of a blower motor inoperative in only one speed.
Answer B is wrong. An intermittent short will affect all blower speeds.
Answer C is wrong. A faulty relay would only affect high-blower speeds.
Answer D is wrong. A loose connector will affect all blower speeds.

Question #30
Answer A is correct. Only Technician A is correct. An in-line filter is sometimes used in the line from the condenser to the evaporator.
Answer B is wrong. Petroleum jelly is not a recommended lubricant for A/C O-rings. The recommended lubricant for all A/C O-rings is mineral oil.
Answer C is wrong. Only Technician A is correct.
Answer D is wrong. Only Technician A is correct.

Question #31
Answer A is wrong. Technician B is also correct.
Answer B is wrong. Technician A is also correct.
Answer C is correct. Both technicians are correct. The tool shown in the figure is a seal protector to be used when replacing the front seal on a compressor. Also, the seal seat O-ring must be installed before the seal.
Answer D is wrong. Both technicians are correct.

Question #32
Answer A is wrong. The screws do not contact the diaphragm.
Answer B is correct. Over-tightening the mounting screws can result in stripped threads. Most HVAC cases are made of plastic so it is important to not use too much force when servicing the components in/on the case.
Answer C is wrong. This will have no effect on the linkage.
Answer D is wrong. A vacuum leak will not likely result from over-torqued screws.

Question #33
Answer A is correct. If the fan clutch is always engaged, it will enhance the operation of the A/C system.
Answer B is wrong. An improperly adjusted blend door cable will cause poor A/C performance.
Answer C is wrong. A low refrigerant charge will cause poor A/C system performance.
Answer D is wrong. A refrigerant overcharge will cause poor A/C system performance.

Question #34
Answer A is correct. A/C fittings should not be disturbed if they are not leaking.
Answer B is wrong. An A/C maintenance service should include cleaning the condenser fins.
Answer C is wrong. An A/C maintenance service should include straightening bent condenser fins.
Answer D is wrong. An A/C maintenance service should include checking all component mounts and insulators.

Question #35
Answer A is wrong. Technician B is also correct.
Answer B is wrong. Technician A is also correct.
Answer C is correct. Both technicians are correct. Orifice tube systems use an accumulator in the suction line to store and dry the refrigerant. TXV systems use a receiver-dryer in the liquid line to store and dry the refrigerant.
Answer D is wrong. Both technicians are correct.

Question #36
Answer A is wrong. The coolant control valve does control the flow of coolant through the heater core.
Answer B is correct. The coolant control valve is not part of the water pump. It is usually located in the inlet heater hose somewhere near the fire wall.
Answer C is wrong. The coolant control valve may be cable operated.
Answer D is wrong. The coolant control valve may be vacuum operated.

Question #37
Answer A is correct. Using a belt tension gauge ensures that the belt is properly adjusted.
Answer B is wrong. This method is not correct and would be excessively tight if used.
Answer C is wrong. This method is not correct because the belt needs to have some static deflection. Using a belt tension gauge ensures the correct tension.
Answer D is wrong. This method could cause component damage due to over-tightening.

Question #38
Answer A is correct. Only Technician A is correct. The service port fittings are different and will easily identify which refrigerant should be used.
Answer B is wrong. R-134a should never be vented to atmosphere.
Answer C is wrong. Only Technician A is correct.
Answer D is wrong. Only Technician A is correct.

Question #39
Answer A is wrong. A/C system pressures will vary with altitude changes.
Answer B is wrong. A/C system pressures will vary with ambient temperature. Higher ambient temperature causes the pressures to rise, especially the high-side pressure.
Answer C is wrong. A/C system pressures vary dramatically with ambient humidity changes.
Answer D is correct. A/C system pressures do not vary with cab humidity. However, outside humidity will change A/C system pressures dramatically.

Question #40
Answer A is wrong. Technician B is also correct.
Answer B is wrong. Technician A is also correct.
Answer C is correct. Both technicians are correct. Some ATC microprocessors are built into the ATC control head, but most have separate controllers that are independent of the control head.
Answer D is wrong. Both technicians are correct.

Question #41
Answer A is correct. Antifreeze raises the boiling point of the coolant, not the conditioner.
Answer B is wrong. The coolant conditioner cartridge filters particulate from the coolant.
Answer C is wrong. The coolant conditioner internally lubricates the cooling system.
Answer D is wrong. The coolant conditioner prevents cavitation corrosion of the cylinder liners.

Question #42
Answer A is wrong. Technician B is also correct.
Answer B is wrong. Technician A is also correct.
Answer C is correct. Both technicians are correct. J1991 sets limits of 15 ppm for moisture and 330 ppm for noncondensables (air).
Answer D is wrong. Both technicians are correct.

Question #43
Answer A is wrong. A short circuit in the blower circuit could cause a blown fuse.
Answer B is wrong. A short circuit in an actuator motor could cause a blown fuse.
Answer C is correct A shorted ECT sensor will cause a code to set, but will not blow a fuse.
Answer D is wrong. A damaged connector could cause a short circuit and blow a fuse.

Question #44
Answer A is wrong. The evaporator is not a cooling system component.
Answer B is wrong. A defective heater valve would have been found during the external inspection
Answer C is wrong. A stuck-open thermostat will cause over-cooling of the engine.
Answer D is correct. A blown head gasket is the most likely cause of an internal engine coolant leak.

Question #45
Answer A is wrong. The engine should not be running when an air line is being removed from the vehicle.
Answer B is wrong. There is no need to drain water from the system before replacing a hose.
Answer C is correct. A technician should drain all air from the system before attempting to replace any chassis air component.
Answer D is wrong. There is no need to remove the compressor from the vehicle when replacing an air hose.

Question #46
Answer A is wrong. Recycled refrigerant has been filtered to remove contaminants and oil.
Answer B is correct. Refrigerant that has been removed and stored is referred to as recovered. Most new A/C machines recycle the refrigerant as the machine recovers the refrigerant from a vehicle.
Answer C is wrong. Reclaimed refrigerant has also been reprocessed to return it to new product quality.
Answer D is wrong. Refining is not a process used with refrigerants.

Question #47
Answer A is wrong. The heater core does not need insulation.
Answer B is wrong. The heater core does not contribute significantly to cab noise.
Answer C is correct. Foam tape can be used to cushion and seal around the heater core. A technician should be very careful when replacing a heater core to make sure that the new core is secure and sealed.
Answer D is wrong. A gasket is provided to seal around heater hose connections.

Question #48
Answer A is wrong. The shape of a preformed hose is determined during the manufacturing process.
Answer B is correct. The spring provides internal support for the lower hose so it will not collapse as the water pump pulls water from the radiator into the engine.
Answer C is wrong. The pressure resilience of the hose comes from the rubber.
Answer D is wrong. A spring inside of a hose does not prevent cavitation, which is caused when air bubbles form and collapse.

Question #49
Answer A is correct. A misaligned air duct could cause inadequate air flow from one or more vents. This problem could also be caused by a foreign object blocking the duct.
Answer B is wrong. A faulty blower resistor will affect air flow from all vents in one or more blower speed settings.
Answer C is wrong. A clogged heater core will not affect output air flow. A clogged core would cause a lack of heat problem.
Answer D is wrong. High humidity will not affect output air flow.

Question #50
Answer A is wrong. The oil level could be verified by flushing the complete system and then adding the specified amount of oil to each component
Answer B is correct. Only Technician B is correct. The refrigeration oil is distributed throughout the system.
Answer C is wrong. Only Technician B is correct.
Answer D is wrong. Only Technician B is correct.

Question #51
Answer A is wrong. A defective clutch will not cycle at all.
Answer B is wrong. A defective control switch will cause the system to be inoperative and have no compressor operation.
Answer C is wrong. An overcharged system will cause the clutch to remain engaged and the compressor operation will be louder than normal due to the excess pressures.
Answer D is correct. Low refrigerant charge will cause the clutch to cycle more frequently than normal.

Question #52
Answer A is wrong. A defective thermostat can cause poor coolant circulation.
Answer B is wrong. An eroded water pump impeller will cause poor coolant circulation.
Answer C is correct. The upper hose is usually under pressure and is unlikely to collapse.
Answer D is wrong. The lower hose is on the suction side of the water pump and could collapse, even though it usually contains a spring to keep this from happening.

Question #53
Answer A is wrong. The cycling switch does not sense outside temperature.
Answer B is correct. The cycling switch is mounted in the accumulator where it senses pressure. The cycling switch usually opens at about 20–25 psi and closes at about 40–45 psi.
Answer C is wrong. The cycling switch senses accumulator pressure, not temperature.
Answer D is wrong. The cycling switch does not sense engine ambient temperature.

Question #54
Answer A is correct. Replacement is the only good repair for an HVAC control panel with an air leak.
Answer B is wrong. This process is not advisable because the head is not designed to be serviced in this way.
Answer C is wrong. This process will not work because grease will not improve the operation of the leaking head.
Answer D is wrong. The selector levers have nothing to do with a leaking control panel.

Question #55
Answer A is wrong. Technician B is also correct.
Answer B is wrong. Technician A is also correct.
Answer C is correct. Both technicians are correct. Insufficient heater output could be caused by a clogged heater core or a misadjusted heater control valve. Either problem could cause a low amount of heated water to pass through the core.
Answer D is wrong. Both technicians are correct.

Question #56
Answer A is wrong. The component shown is an A/C muffler and this will have no effect on A/C circuit pressures.
Answer B is wrong. The component shown is an A/C muffler and this will have no effect on refrigerant filtration.
Answer C is wrong. The component shown is an A/C muffler and this will not affect evaporator icing.
Answer D is correct.The component shown is an A/C muffler and if it was not used, A/C operation would be noisier.

Question #57
Answer A is wrong. A hydrometer is commonly used to measure the specific gravity of coolant but does not provide as accurate a measurement of antifreeze protection as a refractometer.
Answer B is wrong. SCA test strips measure the degree of supplemental coolant additive protection.
Answer C is wrong. Litmus tests measure the acidity/alkalinity of coolant.
Answer D is correct. A refractometer is the recommended instrument for accurately measuring antifreeze protection in a coolant.

Question #58
Answer A is wrong. Most late-model A/C systems are controlled by the engine fuel management system.
Answer B is correct. Only Technician B is correct. Most late-model A/C systems are controlled by the engine control system.
Answer C is wrong. Only Technician B is correct.
Answer D is wrong. Only Technician B is correct.

Question #59
Answer A is wrong. Coolant conditioner should never be added to ELC.
Answer B is wrong. Premixed ELC is already at the correct strength and water should NEVER be added to it.
Answer C is wrong. Never add propylene glycol coolant to a system that has ELC.
Answer D is correct.Nothing should be added to ELC premix since it is already at the correct mixture level.

Question #60
Answer A is wrong. The purchaser is under no legal obligation to maintain records.
Answer B is correct. Only Technician B is correct. Refrigerant suppliers must maintain records about all facilities to which refrigerant is sent. The seller must retain these sales records for three years.
Answer C is wrong. Only Technician B is correct.
Answer D is wrong. Only Technician B is correct.

Question #61
Answer A is wrong. Technician B is also correct.
Answer B is wrong. Technician A is also correct.
Answer C is correct. Both technicians are correct. The resistance of the clutch coil can be tested with an ohmmeter. Another test that can be performed is supplying power and ground to the coil to see if the clutch will engage.
Answer D is wrong. Both technicians are correct.

Question #62
Answer A is wrong. Any acids that form in the A/C system are created due to moisture and corrosion and are usually in a liquid state.
Answer B is correct. Air is not condensable at A/C system pressures. To identify if a vessel of refrigerant has excessive air, a technician can check the pressure in the vessel and compare this pressure with the pressure/temperature chart.
Answer C is wrong. Moisture in the A/C system only evaporates when the system is drawn under a vacuum and is therefore normally in a liquid state.
Answer D is wrong. Refrigeration oil is normally in a liquid state.

Question #63
Answer A is correct. Only Technician A is correct. A cloudy sight glass indicates that the desiccant pack in the receiver/drier has broken.
Answer B is wrong. The receiver/drier must be replaced if the dessicant pack breaks down or if the system has been open for an extended period of time.
Answer C is wrong. Only Technician A is correct.
Answer D is wrong. Only Technician A is correct.

Question #64
Answer A is wrong. Technician B is also correct.
Answer B is wrong. Technician A is also correct.
Answer C is correct. Both technicians are correct. Most ATC systems have some type of internal diagnostic routine. You can also display DTCs on the control panel or a scan tool.
Answer D is wrong. Both technicians are correct.

Question #65
Answer A is wrong. A leaking heater core could cause windshield fogging in the DEFROST mode.
Answer B is wrong. A clogged evaporator drain could cause windshield fogging in the DEFROST mode.
Answer C is wrong. An exterior water leak into the air intake plenum could cause windshield fogging in the DEFROST mode.
Answer D is correct. Moisture in the refrigerant could cause acid to form when it mixes with the refrigerant or it could cause ice to form near the expansion device.

Question #66
Answer A is wrong. A clogged evaporator drain will cause water to leak into the cab or possibly a mildew smell.
Answer B is correct. A cracked evaporator case could cause a whistling noise with the blower motor on high speed.
Answer C is wrong. A broken blend door cable will cause a loss of control of the temperature of the discharge air from the HVAC system.
Answer D is wrong. A low refrigerant charge will cause poor cooling and low system pressures.

Question #67
Answer A is wrong. If the compressor clutch coil were defective, then the clutch would not engage at all.
Answer B is wrong. A defective relay will prevent the clutch from engaging.
Answer C is correct. If the air gap is too large, the clutch may slip briefly upon engagement.
Answer D is wrong. A worn clutch bearing may cause a loud clutch engagement but not a slipping clutch.

Question #68
Answer A is wrong. If the binary switch was open then the system would not operate in the other A/C modes.
Answer B is correct. Only Technician B is correct. The defrost switch contacts could be the cause of the compressor clutch not operating in defrost mode.
Answer C is wrong. Only Technician B is correct.
Answer D is wrong. Only Technician B is correct.

Question #69
Answer A is wrong. Electronic blend door motors are not PWM controlled.
Answer B is correct. Only Technician B is correct. Electronic blend door motors use some sort of feedback device. A potentiometer is typically used to signal the HVAC control device the correct door position.
Answer C is wrong. Only Technician B is correct.
Answer D is wrong. Only Technician B is correct.

Question #70
Answer A is wrong. The relief valve cannot be calibrated.
Answer B is wrong. The relief valve does not need to be replaced if it vents refrigerant to the atmosphere and then resets itself.
Answer C is correct. The valve will reset itself when A/C system pressure returns to a safe level. The valve is designed to release pressure at 450–550 psi.
Answer D is wrong. All mobile A/C systems use a high pressure relief valve.

Question #71
Answer A is wrong. If the technician charges into the high-side with the engine running, high pressure refrigerant may be forced into the container or charging machine.
Answer B is correct. Only Technician B is correct. If liquid refrigerant enters the compressor, then the compressor will be damaged because liquid cannot be compressed.
Answer C is wrong. Only Technician B is correct.
Answer D is wrong. Only Technician B is correct.

Question #72
Answer A is wrong. A refrigerant overcharge will cause high system pressures.
Answer B is wrong. An overheated engine will cause elevated high-side pressure due to high condenser temperature.
Answer C is wrong. A restricted air flow through the condenser will cause elevated high-side pressure due to the lack of heat transfer resulting from restricted air flow.
Answer D is correct. A blockage of the orifice tube screen will cause low high-side pressure.

Question #73
Answer A is correct. R-134a has a faint ether-like odor. However, it is very rare to actually be able to smell a refrigerant leak. A leak of this size would drain the system in a short time.
Answer B is wrong. R-12 is an odorless gas that would not be detectable to a person.
Answer C is wrong. HVAC system input air is not drawn from under the hood or the cab.
Answer D is wrong. A leaking heater core will produce a sweet odor, not an ether-like odor.

Question #74
Answer A is wrong. A plugged evaporator drain may cause windshield fogging, but not an oily film on the windshield.
Answer B is correct. Only Technician B is correct. A refrigerant leak in the evaporator core may allow some refrigerant and oil to leak and cause a thin oily film on the windshield.
Answer C is wrong. Only Technician B is correct.
Answer D is wrong. Only Technician B is correct.

Question #75
Answer A is wrong. Excessive pressure in the system would have to be high enough to rupture a hose, and that is unlikely.
Answer B is correct. Refrigerant leaking from a hose connection will usually contain some refrigeration oil, which quickly collects dirt.
Answer C is wrong. A shaft seal will only cause an oil residue at the front of the compressor.
Answer D is wrong. Too much oil in the system reduces system efficiency and causes low cooling because the oil will coat the heat exchangers.

Question #76
Answer A is wrong. This is an effective method for finding large leaks.
Answer B is wrong. An electronic leak detector is one of the most accurate methods of leak detection.
Answer C is correct. A flame-type leak detector is not recommended. Using a flame-type leak detector with R-12 produces toxic phosgene gas. Flame-type leak detectors do not change colors when exposed to R-134a, so they are not useful.
Answer D is wrong. Using a black-light detector is recommended and effective.

Question #77
Answer A is wrong. Visible coolant flow in the upper radiator tank is an indication that the thermostat has opened.
Answer B is wrong. The upper radiator hose gets hot when the thermostat opens and hot coolant begins flowing to the radiator.
Answer C is correct. The temperature of the lower radiator hose is not a good indicator of an open thermostat.
Answer D is wrong. When the thermostat opens, the temperature should stop rising and stabilize in the normal range.

Question #78
Answer A is wrong. A vacuum leak could cause the HVAC mode switch to not function correctly if the modes are changed using vacuum actuators.
Answer B is wrong. An air leak could cause the HVAC mode switch to not function correctly if the modes are changed using air actuators.
Answer C is wrong. A broken cable could cause the HVAC mode switch to not function correctly if the modes are changed by cable connections.
Answer D is correct. An open blower resistor will not affect mode switch operation. A resistor problem usually affects the blower motor in the low speeds.

Question #79
Answer A is wrong. There are many other tools required to properly diagnose HVAC system failures.
Answer B is correct. Only Technician B is correct. A small thermometer can be used to monitor the performance of an HVAC system. The thermometer should be placed in the center duct during a performance test to show the temperature of the discharge air.
Answer C is wrong. Only Technician B is correct.
Answer D is wrong. Only Technician B is correct.

Question #80
Answer A is correct. Only Technician A is correct. The orifice tube screen is installed to prevent particulate from circulating through the system in the event that the desiccant bag in the accumulator breaks down.
Answer B is wrong. Atomization is not important to the evaporation of refrigerant.
Answer C is wrong. Only Technician A is correct.
Answer D is wrong. Only Technician A is correct.

Question #81
Answer A is wrong. The fan blade will freewheel at a reduced speed due to friction/viscous forces.
Answer B is wrong. The engine idle will rise slightly or be unaffected.
Answer C is wrong. The shutters may or may not be closed.
Answer D is correct. The fan blade may freewheel at a reduced speed. A viscous fan clutch operates by not requiring the fan to turn at engine speed when the temperature is cool. As the fan heats up, the viscous connection speeds up the fan to help cool the engine back down.

Question #82
Answer A is wrong. Oil is not added to the heater core prior to installation.
Answer B is wrong. Oil is not added to the heater core prior to installation.
Answer C is wrong. Neither technician is correct.
Answer D is correct. Neither technician is correct. Both technicians are stating that refrigerant oil needs to be added during heater core replacement and this is not required.

Question #83
Answer A is wrong. Figure A is an R-12 service port fitting. The low side fitting is $\frac{7}{16}$ inch by 20 TPI and the high side is $\frac{3}{8}$ inch by 24 TPI.
Answer B is wrong. Figure B is an R-134a service port fitting. The low side fitting size is 13 mm and the high side fitting size is 16 mm.
Answer C is wrong. Neither technician is correct.
Answer D is correct. Neither technician is correct. Figure A is an R-12 fitting and figure B is an R-134a fitting.

Question #84
Answer A is wrong. In properly maintained cooling systems, clogged heater cores are rare.
Answer B is wrong. Control units are relatively reliable.
Answer C is wrong. Heater control valves do fail, but this is not the most common problem in HVAC systems.
Answer D is correct. Coolant leaks are the most common maintenance problem in truck HVAC systems. A pressure tester is useful in assisting the technician in finding coolant leaks. When pressurizing the system, never apply more than the cap-rated pressure to prevent unwanted component failure.

Question #85
Answer A is correct. If the coolant temperature rises to a predetermined level, the ECM will disengage the compressor clutch. This helps an overheating engine by allowing the condenser to cool down and by removing the mechanical load of driving the compressor.
Answer B is wrong. The IAT sensor signal is not used to control compressor operation.
Answer C is wrong. The oxygen sensor signal is not used to control compressor operation.
Answer D is wrong. There is no cooling fan sensor.

Question #86
Answer A is wrong. The blend door actuator should not bleed vacuum.
Answer B is wrong. A vacuum operated actuator does not contain a motor. The actuator contains a spring that returns it to the normal position.
Answer C is correct. A vacuum actuator should hold vacuum for at least one minute. If the vacuum leaks at all while testing with a vacuum pump, the actuator should be replaced.
Answer D is wrong. The vacuum actuator diaphragm is not porous.

Question #87
Answer A is wrong. Technician B is also correct.
Answer B is wrong. Technician A is also correct.
Answer C is correct. Both technicians are correct. The oil level needs to be checked and corrected following a loss of a large amount of the system oil. It is also very important to use the correct type of oil that is recommended by the manufacturer.
Answer D is wrong. Both technicians are correct.

Question #88
Answer A is wrong. Replacing the ATC control computer should only take place after following the appropriate diagnostic procedure.
Answer B is correct. Only Technician B is correct. A technician should follow and perform the diagnostic steps in the truck manufacturer's service manual.
Answer C is wrong. Only Technician B is correct.
Answer D is wrong. Only Technician B is correct.

Question #89
Answer A is wrong. It is important to keep water drained from the system to reduce internal corrosion.
Answer B is wrong. Some fan clutches are operated using chassis air.
Answer C is wrong. Most radiator shutter systems are air-operated.
Answer D is correct. System air is not used to actuate the engine thermostat. Thermostats are self-contained and internally open and close at their rated temperature to maintain the correct engine temperature.

Question #90
Answer A is correct. Only technician A is correct. The high pressure refrigerant at the compressor outlet can raise the temperature of the outlet to that of the engine coolant or even higher on a hot and humid day.
Answer B is wrong. This is a normal condition caused by the compressor raising the pressure/temperature of the vapor refrigerant.
Answer C is wrong. Only Technician A is correct.
Answer D is wrong. Only Technician A is correct.

Question #91
Answer A is correct. Only Technician A is correct. This is the proper method of checking the oil level in the compressor.
Answer B is wrong. Simply adding an oil charge could result in too much lubricant in the system. This could result in elevated system pressures and lower A/C performance.
Answer C is wrong. Only Technician A is correct.
Answer D is wrong. Only Technician A is correct.

Question #92
Answer A is wrong. A vacuum check valve does not delay the passage of vacuum to components.
Answer B is wrong. A check valve does not switch vacuum on or off.
Answer C is correct. The check valve prevents a vacuum drop during periods of low source vacuum. Low vacuum in a gas engine will occur any time a heavy load is experienced, such as going up a hill or using heavy throttle to pass another vehicle.
Answer D is wrong. A vacuum check valve is not a sensor. Engine vacuum is monitored by the MAP sensor.

Question #93
Answer A is wrong. A defective blower switch could prevent clutch engagement by not allowing current to flow from the white wire to the mode control switch.
Answer B is wrong. An open binary pressure control switch will prevent clutch engagement because it is in the path to the A/C clutch.
Answer C is wrong. A poor connection at the thermostat switch could prevent clutch engagement.
Answer D is correct. If this circuit breaker opens, it only affects the panel light and will not prevent clutch engagement.

Question #94
Answer A is wrong. The condenser is located out in front of the radiator, not under the dash where the sound is coming from.
Answer B is wrong. The mode door does not move when the temperature setting is changed.
Answer C is correct. Bad drive gears in the blend door actuator will cause this symptom. The motor/actuator will have to be replaced to correct this problem.
Answer D is wrong. Arcing in the control head will not cause a grinding noise.

Question #95
Answer A is wrong. Cylinder pressure is not an accurate measure of contents level.
Answer B is correct. The total weight of the cylinder must not exceed the weight of the cylinder when it is empty plus the maximum rated net weight.
Answer C is wrong. Refrigerant cylinders are not equipped with safety relief valves.
Answer D is wrong. Shaking the cylinder is not an accurate measure of contents level.

Question #96
Answer A is wrong. Compression fittings are not used in mobile A/C systems.
Answer B is correct. R134a systems use SAE-approved quick-connect couplings on the service ports. The couplings on the low side of the system are 13 mm in diameter and the couplings on the high side of the system are 16 mm in diameter.
Answer C is wrong. R134a does not use the same fittings as R12 Systems.
Answer D is wrong. 10 mm threaded ports are not used on R134a systems.

Question #97
Answer A is wrong. In LO, current flows through both resistors, so neither one could be open.
Answer B is correct. Only Technician B is correct. The switch provides the path for current to the medium speed resistor. Another possible cause for this problem could be an open in circuit #72, which is the light blue wire leading from the switch to the blower resistor.
Answer C is wrong. Only Technician B is correct.
Answer D is wrong. Only Technician B is correct.

Question #98
Answer A is correct. Only Technician A is correct. Coolant condition should always be tested before replacing a coolant conditioning cartridge to prevent over-conditioning of the coolant.
Answer B is wrong. Routine changing of coolant conditioning/filter cartridges can result in over-conditioned coolant and failure of components in the cooling system.
Answer C is wrong. Only Technician A is correct.
Answer D is wrong. Only Technician A is correct.

Question #99
Answer A is wrong. An open circuit is indicated by an FMI of 5.
Answer B is wrong. A short circuit to ground is indicated by an FMI of 4.
Answer C is correct. This failure code indicates a possible problem with the control module. The technician should follow a diagnostic flow chart to troubleshoot the fault.
Answer D is wrong. A loose connector would probably be indicated by multiple FMI's.

Question #100
Answer A is wrong. The relief valve is a self-resetting valve.
Answer B is wrong. The relief valve can be individually serviced.
Answer C is wrong. Neither technician is correct.
Answer D is correct. Neither technician is correct. The high pressure relief valve resets itself when system pressures return to normal. The high pressure relief valve blows at pressures of 450–550 psi. Also, the high pressure relief valve is individually replaceable.

Question #101
Answer A is wrong. A leaking dash vacuum switch will hiss and cause some control problems but will not cause total loss of vacuum control to the mode doors.
Answer B is wrong. A defective A/C compressor would cause total loss of any cold air.
Answer C is correct. The loss of vacuum supply to the control panel results in a fail-safe mode of all air to the defrost outlets.
Answer D is wrong. A heater control valve failure causes either no cold air or no hot air.

Question #102
Answer A is wrong. The binary switch also provides high pressure protection.
Answer B is wrong. The binary switch also provides low pressure protection.
Answer C is correct. The binary switch disables the A/C compressor when the system pressure is too high or too low. The binary switch is usually mounted on the receiver dryer and is used on a TXV system.
Answer D is wrong. The binary switch does provide protection for the compressor in low or high pressure situations.

Question #103
Answer A is correct. Only Technician A is correct. If the expansion valve is stuck open, the low-side pressure will be high and the compressor will run continuously.
Answer B is wrong. A defective compressor clutch would cause the pressures to be equal because the compressor would not engage.
Answer C is wrong. Only Technician A is correct.
Answer D is wrong. Only Technician A is correct.

Question #104
Answer A is wrong. The orifice tube is located in the evaporator core inlet.
Answer B is correct. The orifice tube allows low pressure liquid to be metered into the evaporator. Since the pressure is much lower after the orifice tube, the low pressure refrigerant starts to boil and take on heat from the surrounding surface and air.
Answer C is wrong. The orifice tube does not affect the refrigerant flow through the condenser.
Answer D is wrong. The orifice tube does not affect air flow through the evaporator.

Question #105
Answer A is wrong. Thermostats can be tested by monitoring engine temperature during the warm-up process to make sure the thermostat does not open too early. The thermostat can also be tested by removing it from the truck and submerging it in hot water to see when it opens.
Answer B is correct. A technician uses a pressure tester to test radiators, pressure caps, and hoses for leaks. Caution should be taken to not apply pressure higher than the radiator cap is rated.
Answer C is wrong. A/C leaks must be located using a leak detector.
Answer D is wrong. Vacuum diaphragms are tested using a hand-held vacuum pump.

Question #106
Answer A is wrong. Twenty minutes is not enough time to evacuate.
Answer B is wrong. Ten minutes is not enough time to evacuate.
Answer C is wrong. Fifteen minutes is not enough time to evacuate.
Answer D is correct. To ensure that the entire system is under deep enough vacuum to remove all moisture, the pump should be run for at least thirty minutes. After evacuating for thirty minutes, the technician should let the system sit with the vacuum pump turned off and make sure that the vacuum is maintained. If the system does hold a vacuum, then there are no major leaks in the system and it can be charged with refrigerant.

Question #107
Answer A is wrong. Some coolant control valves close when vacuum is removed.
Answer B is correct. Only Technician B is correct. This process allows the technician to check the specific coolant flow valve operation. The valve should hold the vacuum for at least one minute without leaking down.
Answer C is wrong. Only Technician B is correct.
Answer D is wrong. Only Technician B is correct.

Question #108
Answer A is correct. A compressor with worn reed valves would be unable to create very much pressure difference between high- and low-side readings.
Answer B is wrong. No pressure would most likely mean the system is empty.
Answer C is wrong. Both gauges reading too low would likely indicate an undercharge, or a restriction.
Answer D is wrong. When both gauges read too high, it could be an overcharge, high operating temperatures, air or moisture in the system, or a TXV valve stuck open.

Question #109
Answer A is wrong. A stuck water-control valve controls temperature, not mode selection.
Answer B is correct. Small objects in the ducts can block the movement of the mode door. Many times these object enter the HVAC duct by falling down the defroster grill or by overfilling the glove box.
Answer C is wrong. There would be no air flow from the outlets if the blower motor were inoperative.
Answer D is wrong. The blend door changes the temperature of the air flow, not its outlet.

Question #110
Answer A is wrong. Technician B is also correct.
Answer B is wrong. Technician A is also correct.
Answer C is correct. Both technicians are correct. If moisture enters the A/C system, it mixes with the refrigerant to form harmful acids. The method of removing any moisture from the A/C system is to pull a vacuum on the system for at least 30 minutes. Pulling a vacuum on the system will cause any moisture to vaporize and be pulled out of the system.
Answer D is wrong. Both technicians are correct.

Question #111
Answer A is wrong. The high pressure switch does not provide a boosting function.
Answer B is correct. The A/C high pressure switch opens the electrical circuit to the compressor clutch coil when high-side pressure reaches its upper limit. The high pressure switch typically opens at about 380–420 psi.
Answer C is wrong. The high pressure switch does not maintain pressure.
Answer D is wrong. The high pressure switch does perform a venting function.

Question #112
Answer A is wrong. Some cabin air filters are made of paper or a metallic mesh.
Answer B is wrong. Cabin air filters are not designed to remove moisture from the air.
Answer C is wrong. Cabin air filters are replaceable and should be checked during routine maintenance checks.
Answer D is correct. Cabin air filters are designed to remove dust and dirt from cab and sleeper air. These filters should be inspected while performing a preventative maintenance inspection and should be replaced if found to be dirty or restricted.

Question #113
Answer A is wrong. The microprocessor does not control the modes on a semiautomatic HVAC system. The processor controls only the temperature in these systems.
Answer B is wrong. A bad blend door actuator would only affect the temperature of the air.
Answer C is wrong. A blower resistor problem would only affect the blower speeds.
Answer D is correct. An improperly adjusted mode door cable could cause the air not to be directed to the desired location.

Question #114
Answer A is wrong. Technician B is also correct.
Answer B is wrong. Technician A is also correct.
Answer C is correct. Both technicians are correct. There is a definite problem with this vehicle's ATC system. The in-cab temperature sensor could be out of calibration and providing a false input to the processor. This problem could also be caused by a sticking blend door.
Answer D is wrong. Both technicians are correct.

Question #115
Answer A is correct. Technician A is correct. Refrigerant enters the compressor as a low pressure gas. The compressor raises the pressure and temperature of the refrigerant and it leaves the compressor as a hot/high pressure gas through the discharge line.
Answer B is wrong. Refrigerant changes state from a gas to liquid; that is, it "condenses" in the condenser. Boiling is the change of state from a liquid to a gas.
Answer C is wrong. Only Technician A is correct.
Answer D is wrong. Only Technician A is correct.

Question #116
Answer A is wrong. A defective coolant control valve could cause this problem by not regulating coolant flow into the heater core properly.
Answer B is correct. A defective engine coolant temperature sensor should not cause a problem with temperature control system.
Answer C is wrong. A defective air control solenoid could cause this problem by not changing the blend door the correct way.
Answer D is wrong. A defective blend air door air cylinder could cause this problem.

Question #117
Answer A is wrong. An overheating engine will result in the exact opposite condition; i.e., a lean condition.
Answer B is wrong. A defective radiator cap will cause engine overheating.
Answer C is correct. With the engine thermostat stuck open, the engine will not reach operating temperature. This will result in the engine coolant sensor sending a cold engine message to the computer, resulting in a rich air–fuel mix.
Answer D is wrong. A stuck open coolant valve only affects truck interior heating and cooling.

Question #118
Answer A is correct. If the expansion valve is stuck closed, the low side of the system will be pulled down into a vacuum and the compressor will not be able to sufficiently pressurize the high side.
Answer B is wrong. Without flow from the evaporator, the compressor cannot generate high pressure.
Answer C is wrong. With the low side being restricted, the low-side pressure will be nearing a vacuum.
Answer D is wrong. This condition would be caused by a faulty compressor.

Question #119
Answer A is wrong. Shut-off valves do not have to be closed every time the A/C system is switched off. However, they must be closed when connecting and disconnecting the hoses from the vehicle.
Answer B is correct. Only Technician B is correct. New environmental laws dictate that shut-off valves must be located no more than 12 inches from test hose service end.
Answer C is wrong. Only Technician B is correct.
Answer D is wrong. Only Technician B is correct.

Question #120
Answer A is correct. It is the lowest value in the choices and represents the maximum voltage drop allowed across a low amperage ground circuit.
Answer B is wrong. The drop is too large, indicating high resistance in the ground circuit.
Answer C is wrong. The drop is too large, indicating high resistance in the ground circuit. This number represents the maximum voltage drop in a battery cable with the engine cranking.
Answer D is wrong. The drop is too large, indicating high resistance in the ground circuit.

Question #121
Answer A is correct. Only Technician A is correct. A faulty reed valve could cause the high-side pressure to be low and the low-side pressure to be high.
Answer B is wrong. An overcharge of refrigerant oil will not cause these symptoms but will reduce the cooling capabilities of the system. The likely pressures that would be present on a system with too much oil is high pressures in both sides of the system.
Answer C is wrong. Only Technician A is correct.
Answer D is wrong. Only Technician A is correct.

Question #122
Answer A is correct. Comparing the pressure of recovered refrigerant to the theoretical pressure of pure refrigerant at a given temperature is the best method of testing for noncondensable gases in refrigerant.
Answer B is wrong. Pressure cannot be compared to humidity.
Answer C is wrong. Pressure cannot be compared to volume.
Answer D is wrong. A halogen leak detector cannot be used to check for noncondensable gases.

Question #123
Answer A is wrong. A faint hissing noise after shutting off the engine is caused by the equalization of refrigerant pressure in the A/C system.
Answer B is wrong. A defective TXV would cause poor cooling accompanied by irregular pressures.
Answer C is wrong. Neither technician is correct.
Answer D is correct. Neither technician is correct. It is normal to have an audible noise in the A/C system after shutting the engine off. The noise comes from the equalizing pressures from each side of the system.

Question #124
Answer A is wrong. Original refrigerant containers should not be used to store recycled refrigerant.
Answer B is wrong. If the valve is simply opened, any remaining refrigerant will be vented to the atmosphere.
Answer C is wrong. There is no reason to introduce oil into the cylinder.
Answer D is correct. After any remaining refrigerant is recovered, the cylinder should be evacuated, marked, and recycled for scrap metal.

Question #125
Answer A is correct. The check valve is a one-way valve that allows the reservoir to hold constant vacuum regardless of fluctuations in the vacuum source.
Answer B is wrong. A defective vacuum pump would cause this condition constantly, not just on shut down and going uphill.
Answer C is wrong. Defective fuel injectors may affect engine operation under load, but not during shutdown.
Answer D is wrong. A leaking actuator diaphragm would cause this condition all of the time.

Question #126
Answer A is wrong. Some actuator motors can be calibrated by putting the control head in self-diagnostic mode.
Answer B is correct. A/C DTCs represent a fault in a specific system, not a component.
Answer C is wrong. The control rods do have to be manually adjusted on a few A/C systems.
Answer D is wrong. The only time a technician has to adjust the motor control rods is after motor replacement or adjustment.

Question #127
Answer A is wrong. In a strategy-based diagnostic process, a technician must check for visual signs before replacing any parts.
Answer B is correct. Only Technician B is correct. The first process is to visually check the components. A technician should also follow the DTC troubleshooting chart for the proper diagnostic sequence.
Answer C is wrong. Only Technician B is correct.
Answer D is wrong. Only Technician B is correct.

Question #128
Answer A is wrong. The blend air type HVAC system modulates temperature control by mixing the flow of air into the cab by using a blend air door. The blend air door regulates the air temperature by routing more air past the heater core if more heat is needed.
Answer B is wrong. ATC systems control all aspects of the HVAC system with a single input from the driver.
Answer C is correct. Chilling and heating water and then re-circulating it are only used in stationary HVAC systems.
Answer D is wrong. SATC systems require the driver to choose the mode and blower speed while the processor directs the position of the temperature door.

Question #129

Answer A is correct. Only Technician A is correct. The best way to repair wiring is by soldering. Heat shrink wrap should also be used to keep out moisture and prevent future corrosion problems. One other quality wire repair technique is using crimp and seal connectors. These connectors also keep out moisture and prevent future corrosion problems.

Answer B is wrong. Twisting wires together and taping them will result in a high resistance connection and ultimately circuit failure.

Answer C is wrong. Only Technician A is correct.

Answer D is wrong. Only Technician A is correct.

Question #130

Answer A is wrong. When the cooling system pressure is increased, the boiling point increases. For every 1 psi of pressure that is added to the cooling system, the boiling point is raised 3°F.

Answer B is correct. When more antifreeze is added to the coolant, the boiling point increases.

Answer C is wrong. A good quality ethylene glycol antifreeze contains antirust inhibitors.

Answer D is wrong. Coolant solutions must be recovered, recycled, and handled as hazardous waste.

Question #131

Answer A is correct. This symptom is typical of an A/C system with an evaporator icing problem, which could be caused by a defective thermal expansion valve.

Answer B is wrong. A defective coolant control valve would not cause the A/C system output air to change from cold to warm.

Answer C is wrong. The blend air door does not typically have a return spring.

Answer D is wrong. The fresh air door does not control the output air temperature. This door has two positions that include fresh and recirculated air.

Question #132

Answer A is wrong. A/C pressure gauges are used to conduct a performance test of the refrigeration system, not to test the accuracy of the ATC system.

Answer B is wrong. ATC system calibration is sometimes not adjustable in the field.

Answer C is wrong. Neither technician is correct.

Answer D is correct. Neither technician is correct. A scan tool is used to recalibrate the ATC controller with the actuators on some systems. Other ATC systems allow the technician to push a series of buttons on the ATC control panel and enter diagnostic and calibration mode. The reason Technician B is wrong is the word **all** in his statement.

Question #133

Answer A is wrong. Using an ohmmeter will typically not provide calibration.

Answer B is wrong. This process only checks that sensor.

Answer C is wrong. Not all ATC units have such a dial.

Answer D is correct. The easiest way to check the calibration of an ATC system is to measure the cab temperature using a thermometer. Some ATC systems can be calibrated by pushing a series of buttons on the ATC control head. This puts the system into diagnostic and calibration mode. Also, some systems are calibrated using scan tools.

Question #134

Answer A is wrong. A technician can test control panel vacuum systems by applying vacuum with a hand pump to the input end of the system, not output.

Answer B is correct. Only Technician B is correct. To test an actuator, a technician should connect the vacuum pump to each vacuum actuator and supply 15 to 20 in. Hg to the actuator and the gauge should stay at a steady vacuum for one minute.

Answer C is wrong. Only Technician B is correct.

Answer D is wrong. Only Technician B is correct.

Question #135
Answer A is correct. The valve is front-seated when the stem is turned clockwise to seat the front valve face to the left. This shuts off the flow of refrigerant to the compressor and isolates it.
Answer B is wrong. Back seating the valve moves it counterclockwise to the right to seal the rear face valve. This is the normal valve position.
Answer C is wrong. When the valve is at mid-position, it is not seated at all. The valve should be in this position only after connecting the hose for the manifold gage set. It remains in this position while conducting any service to the system but must be back-seated before removing the service hose.
Answer D is wrong. In normal operating position, the valve is back-seated.

Question #136
Answer A is correct. The thermal bulb must be in contact with the evaporator outlet tube. The thermal bulb serves as the temperature-sensing part of the TXV. If the bulb senses warm temperatures, the TXV opens to allow more refrigerant to enter the evaporator core. If the bulb senses cold temperatures, the TXV closes to restrict the flow of refrigerant into the evaporator core.
Answer B is wrong. This area is the wrong location. The condenser is in the high side of the system.
Answer C is wrong. This is close to where the thermal bulb is, but the bulb is located inside the HVAC duct attached to the evaporator outlet line.
Answer D is wrong. The thermal bulb does not directly contact the refrigerant.

Question #137
Answer A is wrong. A technician completes the charging process when the correct weight of refrigerant has entered the system.
Answer B is wrong. If the low side does not move from a vacuum to a pressure, there is a restriction.
Answer C is wrong. The truck engine does not need to be running when recharging with a recharging machine.
Answer D is correct. OEM's recommend either a high-side (liquid) or a low-side (vapor) charging process.

Question #138
Answer A is wrong. Moisture cannot be removed from the A/C system after charging the system with new refrigerant. However, all systems have a drying device that should capture any trace amounts of moisture that enters the system.
Answer B is correct. Only Technician B is correct. Moisture that enters the A/C system will be harmful to the system and cause poor performance because of internal erosion and the possibility of the moisture turning to ice near the expansion device.
Answer C is wrong. Only Technician B is correct.
Answer D is wrong. Only Technician B is correct.

Question #139
Answer A is wrong. It is advisable to remove the negative battery cable before control panel service.
Answer B is correct. A technician does not have to recover the refrigerant before removing the control panel.
Answer C is wrong. If the truck contains a supplemental restraint system, wait the specified period after you remove the negative battery cable.
Answer D is wrong. Self-diagnostic tests may indicate a defective control panel in an ATC system.

Question #140
Answer A is wrong. There is no need to remove control cables from the vehicle before replacing any electronic control panel or device that could have solid state circuitry.
Answer B is correct. The technician should disconnect the batteries before replacing an electronic control panel. This will help prevent a voltage spike when handling the unit. The technician should also ground himself before touching the electronic panel to prevent static electricity from damaging it.
Answer C is wrong. Some HVAC electronic control panels can be replaced without disassembling the dash panels.
Answer D is wrong. Nothing should be applied to the switch contacts.

Question #141
Answer A is wrong. A problem that causes the shutters to remain closed with the A/C on could cause very high system pressures and trip high pressure relief valve.
Answer B is wrong. A clogged condenser could cause the relief valve to trip by causing excessive high-side pressure.
Answer C is wrong. An inoperative fan clutch could cause the relief valve to trip by not moving enough air across the condenser.
Answer D is correct. A defective A/C compressor will not cause the relief valve to trip. If the compressor were defective, it would not be able to develop the high pressure needed to trip the pressure relief valve. The relief valve typically opens up at approximately 450 to 550 psi.

Question #142
Answer A is wrong. To prevent damaging A/C system components, the nitrogen pressure must be regulated.
Answer B is wrong. The A/C compressor must be disconnected before flushing the system to prevent debris from entering into it.
Answer C is wrong. To avoid damaging restrictive components, they must be removed before flushing.
Answer D is correct. It is very acceptable to vent nitrogen to the atmosphere. The atmosphere already contains an abundant amount of nitrogen, so no harm will be done by venting nitrogen during A/C service.

Question #143
Answer A is wrong. Technician B is also correct.
Answer B is wrong. Technician A is also correct.
Answer C is correct. Both technicians are correct. Engine protection systems will first derate the engine and then kill the engine if critical situations develop in the engine such as low oil pressure or high coolant temperature. Most protection systems will allow the truck to be restarted and moved in case the truck is in a dangerous location such as on a train track.
Answer D is wrong. Both technicians are correct.

Question #144
Answer A is correct. A 50/50 mix will result in a very slow windshield defrost.
Answer B is wrong. 85 percent of the output air directed to the windshield in the DEFROST mode is an appropriate mix.
Answer C is wrong. A 75 percent air flow to the windshield will defrost the windshield better than a 50 percent air flow to the windshield.
Answer D is wrong. A 70 percent air flow to the windshield will defrost the windshield better than a 50 percent air flow to the windshield.

Question #145
Answer A is correct. Refrigerant returns to the compressor as a low pressure gas. This line is known as the suction line and is very cold on a normally operating system.
Answer B is wrong. Refrigerant in an operating A/C system is a low pressure liquid after it passes through the metering device. While in the evaporator, the refrigerant changes from a low pressure liquid into a low pressure gas.
Answer C is wrong. Refrigerant leaves the compressor as a high pressure gas.
Answer D is wrong. Refrigerant in the liquid line is a high pressure liquid. The liquid line connects the condenser to the evaporator. The condenser changes high pressure vapor into a high pressure liquid by allowing the refrigerant to give up heat and condense into a liquid.

Question #146
Answer A is wrong. The low pressure cut-outswitch is not necessarily at fault.
Answer B is correct. Only Technician B is correct. A low refrigerant charge will cause this symptom. The low pressure cut-out switch opens its contacts when the pressure drops below 20–25 psi. The switch acts as a deicing device when the system is operating. The switch also serves as a low refrigerant protection device that keeps the compressor from turning on when the refrigerant gets low.
Answer C is wrong. Only Technician B is correct.
Answer D is wrong. Only Technician B is correct.

Question #147
Answer A is wrong. The condenser should not be replaced, the fins should be straightened, and then it should be cleaned.
Answer B is correct. Only technician B is correct. Several bent fins and a moderate accumulation of dead insects will restrict air flow through the condenser enough to cause poor cooling and higher-than-normal system pressures. It is a good practice to check for condenser restrictions while performing a good preventative maintenance inspection.
Answer C is wrong. Only Technician B is correct.
Answer D is wrong. Only Technician B is correct.

Question #148
Answer A is wrong. The service valve is front seated and this blocks the gauge port so no pressure would register on a manifold gauge set. A technician uses the service valve mid-position when servicing the A/C system.
Answer B is wrong. The system could not operate normally because the passage to the compressor is blocked. Some system pressure may appear at the gauge port.
Answer C is correct. Front seating a service valve isolates the compressor from the system for service.
Answer D is wrong. Pressure from the compressor is blocked. A technician uses the service valve mid-position for service.

Question #149
Answer A is wrong. In-line filters are not a substitute for flushing or replacing components that have been contaminated by debris.
Answer B is wrong. If an in-line filter contains an orifice, then the original orifice must be removed.
Answer C is wrong. Neither technician is correct.
Answer D is correct. Neither technician is correct. In-line filters are sometimes recommended after flushing and/or component replacement as a method to prevent debris from entering into the compressor. If an in-line filter is installed that contains an orifice tube, then the original orifice tube must be removed.

Question #150
Answer A is wrong. The refrigerant does not need to be recovered from the system before the valve is replaced.
Answer B is correct. The heater hoses connected to the coolant control valve may be clamped during the replacement procedure to maintain coolant in the system. Many tool companies sell a tool to perform this function called hose pinching pliers.
Answer C is wrong. The coolant control valve is located in front of the firewall.
Answer D is wrong. The coolant control valve is an individually replaceable component.

Question #151

Answer A is correct. The compressor discharge valve opens after the pressure compresses vaporized refrigerant, allowing the refrigerant to move to the condenser to change from a high pressure gas into a high pressure liquid.

Answer B is wrong. If the discharge valve opened before vaporized refrigerant compressed, it would not properly raise the refrigerant pressure to allow a change from gas to a liquid while in the condenser. In addition, it does not go from the compressor to the evaporator.

Answer C is wrong. The reed valves just open and close; they do not regulate a variable pressure.

Answer D is wrong. Reed valves cannot sense A/C system temperature, so regulation can not take place.

Question #152

Answer A is wrong. The question is asking you to look for a defect that has not caused the problem. If the water control valve were seized, it would not move on command.

Answer B is also wrong. A plugged vacuum hose between the solenoid and valve could result in a no water valve movement.

Answer C is wrong. A jammed actuator linkage could result in no water valve movement on command.

Answer D is correct. The audible click indicated that the solenoid plunger moved so a seized solenoid could not be the cause of the problem.

Question #153

Answer A is wrong. The expansion valve is located near the evaporator, not the condenser.

Answer B is wrong. An accumulator is used with a CCOT system that does use a thermal bulb or capillary tube.

Answer C is correct. The thermal bulb and capillary tube are part of the thermal expansion valve located near the evaporator inside of the duct housing. The thermal bulb and capillary tube sense the temperature of the outlet line of the evaporator core. When the thermal bulb senses warm temperatures, the TXV opens to allow more refrigerant into the evaporator core. When the thermal bulb senses cold temperatures, the TXV closes to restrict refrigerant flow into the evaporator.

Answer D is wrong. A CCOT system does use a thermal bulb or capillary tube.

Question #154

Answer A is wrong. The A/C system does not need to be evacuated for 30 minutes prior to checking the lubricant. The refrigerant does need to be recovered with an approved recovery machine.

Answer B is wrong. The compressor needs to be operated just prior to recovering the refrigerant to ensure that the lubricant is properly distributed through the system.

Answer C is wrong. Modern compressors do not have a dipstick to allow the oil to be checked.

Answer D is correct. The old lubricant should be measured before new lubricant is added. After adding the new lubricant, the compressor should be turned several revolutions by hand to help avoid compressor damage when it is first turned on.

Question #155

Answer A is wrong. SCA stands for supplemental coolant additive, which is a corrosion inhibitor additive. Adding more antifreeze does not change the SCA reading.

Answer B is wrong. Continuing to run the truck until the next PMI will not fix the problem. It is possible if the coolant was only SLIGHTLY over-conditioned the truck could be run until the next service interval and retested. However running a diesel engine with a severely over conditioned cooling system can cause engine damage.

Answer C is correct. The technician needs to drain the entire coolant system and add the proper SCA mixture to achieve the correct percentage of SCA.

Answer D is wrong. Action should be taken to correct the problem as soon as possible.

Question #156
Answer A is wrong. Technician B is also correct.
Answer B is wrong. Technician A is also correct.
Answer C is correct. Both technicians are correct. When evacuating a system, the vacuum pump must run at least 30 minutes to remove moisture from the system. After 5 minutes of pump operation, the low-side gauge should indicate 20 in. Hg, and the high-side gauge should read below zero (unless the needle is restricted by a stop pin). If the high-side gauge does not drop below zero, there may be a refrigerant blockage. Within 15 minutes, the low-side gauge should drop to about 24 to 26 in. Hg. If it doesn't, close the low-side gauge valve and observe the gauge. If it slowly rises, there may be a leak.
Answer D is wrong. Both technicians are correct.

Question #157
Answer A is wrong. A binding door will not cause the actuator to hunt for the proper position. This problem would cause a popping noise in the dash and poor operation of the binding door.
Answer B is wrong. A faulty motor will not cause the actuator to hunt for the desired position. The system would have no change in temperature if the blend door actuator was bad.
Answer C is wrong. Most ATC actuators do not use a cable to connect to the door. Typically, the actuator directly attaches to the door and the feedback sensor is located inside the actuator.
Answer D is correct. A faulty feedback device can cause the blend air door to hunt for the desired position. The feedback device is typically a potentiometer, which uses three wires. One wire is the input voltage from the controller, a second wire is the return ground from the controller, and the third wire is the signal wire to the controller.

Glossary

Absolute Pressure The zero point from which pressure is measured.

Actuator A device that delivers motion in response to an electrical signal.

A/D Converter Abbreviation for analog-to-digital converter.

Additive An additive intended to improve a certain characteristic of a material or fluid.

After-Cooler A heat exchanger often cooled by engine coolant.

Air Conditioning (A/C) The control of air movement, humidity, and temperature by thermodynamics.

Air Dryer Unit that removes moisture.

Ambient Temperature Temperature of the surrounding or prevailing air. The temperature in the service area where testing is taking place.

Amp Abbreviation for ampere.

Ampere The unit for measuring electrical current.

Analog Signal A voltage signal that varies within a given range (from high to low, including all points in between).

Analog-to-Digital Converter (A/D Converter) Device that converts analog voltage signals to a digital format.

Analog Volt/Ohmmeter (AVOM) Test meter used for checking voltage and resistance. Analog meters should not be used on solid state circuits.

Anticorrosion A chemical used to protect metal surfaces from corrosion.

Antifreeze A compound, such as ethylene or propylene glycol added to water to lower its freezing point.

Antirust Agent Additive used with lubricating oils and coolants to prevent rusting of metals.

Armature Rotating component of a (1) starter or other motor, (2) generator, (3) compressor clutch.

ASE Abbreviation for Automotive Service Excellence, a trademark of the National Institute for Automotive Service Excellence.

Atmospheric Pressure The weight of the air at sea level; 14.696 pounds per square inch (psi) or 101.33 kilopascals (kPa).

Axis of Rotation Center line around which a gear or shaft revolves.

Battery Terminal A tapered post or threaded studs on the battery case for connecting cables.

Bimetallic Dissimilar metals joined together that have different bending characteristics when subjected to changes of temperature.

Blade Fuse A type of fuse having two flat male terminals for insertion in connectors.

Block-Diagnosis Chart A troubleshooting chart that lists symptoms, possible causes, and probable remedies in columns.

Blower Fan A fan that blows air through a ventilation, heater, or air conditioning system.

Boss Bearing race.

British Thermal Unit (Btu) A measure of heat quantity. Amount of heat required to raise 1 pound of water 1°F.

Btu Abbreviation for British thermal unit.

CAA Abbreviation for Clean Air Act.

Cartridge Fuse Fuse having a strip of low melting point metal enclosed in a glass tube. If excessive current flows through the circuit, the fuse element melts, opening the circuit.

Cavitation An erosion condition caused by vapor bubble collapse.

CFC Abbreviation for chlorofluorocarbon.

Check Valve Valve that allows fluid to flow in one direction only.

Chlorofluorocarbon (CFC) A compound used in the production of some refrigerants that damage the ozone layer.

Circuit The complete path of an electrical current, including the generating device. When the path is unbroken, the circuit is closed and current flows. When the circuit continuity is broken, the circuit is open and current flow stops.

Clean Air Act (CAA) Federal regulation, passed in 1992, that resulted in major changes in air conditioning systems.

COE Abbreviation for cab-over-engine.

Coefficient of Friction Amount of friction developed between two objects in physical contact when one is drawn across the other.

Compression Applying pressure to a spring or component attempting to reduce its length in the direction of the compressing force.

Compressor Component of an air conditioning system that compresses low pressure refrigerant vapor and pumps it through the refrigeration circuit.

Condensation Process by which gas changes state to a liquid.

Condenser Component in an air conditioning system used to cool refrigerant below its boiling point, changing it from a vapor to a liquid.

Conductor Any material that permits the electrical current to flow.

Coolant Liquid that circulates in an engine cooling system, usually a solution of antifreeze and water.

Coolant Heater A component used to aid engine starting and reduce the wear caused by cold starting.

Coolant Hydrometer A tester designed to measure coolant-specific gravity and to determine the amount of antifreeze protection.

Cooling System Engine system for circulating coolant.

Crankcase The housing within which a crankshaft rotates.

Cycling Repeated on-off action of the air conditioner compressor.

Dampen Reduce oscillation or movement.

Data Links Connection points through which computers communicate with other electronic devices such as control panels, modules, sensors, or other computers.

Deadline Take a vehicle out of service.

Deburring Remove sharp edges from a cut.

Deflection Bending movement as the result of an external force.

Department of Transportation (DOT) A government agency that establishes vehicle standards.

DER Abbreviation for Department of Environmental Resources.

Detergent Additive An additive that helps keep metal surfaces clean and prevents deposits. These additives suspend particles of carbon and oxidized oil in the oil.

Diagnostic Flow Chart Chart that provides a systematic approach to component troubleshooting and repair. Found in service manuals and are vehicle make and model specific.

Dial Caliper A measuring instrument capable of taking inside, outside, depth, and step measurements.

Digital Binary Signal A signal that has two values: on and off.

Digital Diagnostic Reader (DDR) An electronic service tool usually handheld or PC-based.

Digital Multimeter (DDM) Required to diagnose circuits with solid state components.

Digital Volt/Ohmmeter (DVOM) Test meter recommended by most manufacturers for use on solid state circuits.

Diode The simplest semiconductor device formed by joining P-type semiconductor material with N-type semiconductor material. A diode allows current flow in one direction, but not in the opposite direction.

Dispatch Sheet Form used to track dates when the work is to be completed. Some dispatch sheets follow through each step of the servicing process.

Dog Tracking Off-center tracking of the rear wheels as related to the front wheels.

DOT Abbreviation for Department of Transportation.

Driver's Manual Publication that contains information needed by the driver to understand, operate, and care for the vehicle.

ECM Abbreviation for electronic control module. A controller module that contains the CPU and output generators.

ECU Abbreviation for electronic control unit.

EG Ethylene Glycol Commonly used antifreeze agent.

ELC Extended Life Coolant Premixed coolant solution with high resistance to breakdown.

Electricity The movement of electrons from one place to another.

Electromotive Force (EMF) The force that moves electrons between atoms. Measured in units called volts.

Electronic Control Unit (ECU) Module that processes data and generates outcomes in an electronic circuit.

Electronics The technology of controlling electricity.

Electrons Negatively charged particles orbiting a nucleus.

EMF Abbreviation for electromotive force.

Environmental Protection Agency (EPA) An agency of the U.S. government with the responsibilities of protecting the environment and enforcing the Clean Air Act (CAA) of 1990.

EPA Abbreviation for the Environmental Protection Agency.

Evaporator Component in an air conditioning system used to remove heat from air forced through it.

Fatigue Failures Progressive destruction of a shaft or gear teeth material usually caused by overloading.

Fault Code A trouble code that is written to computer memory. Can be read by blink-out or DDR.

Federal Motor Vehicle Safety Standard (FMVSS) A federal standard that specifies that all vehicles in the United States be assigned a Vehicle Identification Number (VIN) and adhere to a range of safety standards.

FHWA Abbreviation for Federal Highway Administration.

Fixed-Value Resistor An electrical device designed to have one resistance rating used for controlling voltage.

Flammable A term used to describe any material that will catch fire or explode.

Flare To spread gradually outward in a bell shape.

FMVSS Abbreviation for Federal Motor Vehicle Safety Standard.

Foot-Pound English unit of measurement for torque.

Franchised Dealership Dealer that has a contract with a manufacturer to sell and service a particular line of vehicles.

Fretting A result of vibration in which the bearing outer race picks up the machining pattern.

Fusible Link A term used for fuse link.

Fuse Link Short length of smaller gauge wire installed in a conductor, usually close to the power source.

GCW Abbreviation for gross combination weight.

Gear A disc-like wheel with external or internal teeth that transmits rotary motion.

Gear Pitch The number of teeth per unit of pitch diameter, an important factor in gear design and operation.

Gross Combination Weight (GCW) Total vehicle weight including payload, fuel, and driver.

Gross Vehicle Weight (GVW) The total weight of a fully equipped vehicle and its payload.

Ground Negatively charged side of an electrical circuit. A ground can be a wire, the negative side of the battery, or the vehicle chassis.

Grounded Circuit A shorted circuit causes a current to return to the battery before it has reached its intended destination.

GVW Abbreviation for gross vehicle weight.

Hazardous Materials Any substance that is flammable, explosive, or known to produce adverse health effects in people or the environment.

Heads-Up Display (HUD) A technology used in some vehicles that superimposes data on the driver's field of vision in the windshield. The operator can view the information, which appears to "float" just above the hood. This allows the driver to monitor conditions such as road speed without interrupting his normal view of traffic.

Heater-Control Valve A valve that controls the flow of coolant into the heater core from the engine.

Heat Exchanger Device used to transfer heat, such as a radiator or condenser.

Heavy-Duty Truck A truck that has a GVW of 26,001 pounds or more.

High-Resistant Circuits Circuits that have an increase in circuit resistance, with a corresponding decrease in current.

Hinged-Pawl Switch The simplest type of switch; one that makes or breaks the current of a single conductor.

HUD Abbreviation for heads-up display.

Hydrometer A tester designed to measure the specific gravity of a liquid.

Inboard Toward the centerline of the vehicle.

In-Line Fuse A fuse in series with the circuit usually enclosed in a small plastic fuse holder. Used as a protection device for a portion of a circuit.

Insulator A material, such as rubber or glass, that offers high resistance to the flow of electrons.

Integrated Circuit A solid state component containing diodes, transistors, resistors, capacitors, and other electronic components mounted on a single piece of material.

Jumper Wire A wire used to temporarily bypass a circuit or components for electrical testing. A jumper wire consists of a length of wire with an alligator clip at each end.

Kinetic Energy Energy in motion.

Lateral Runout Wobble or side-to-side movement of a rotating wheel or shaft.

Linkage System of rods and levers used to transmit motion or force.

Low-Maintenance Battery A conventionally vented, lead/acid battery, requiring periodic maintenance.

LTL Abbreviation for less than truckload.

Magnetorque An electromagnetic clutch.

Maintenance-Free Battery A battery that does not require the addition of water during normal service life.

Maintenance Manual A publication containing routine maintenance procedures for vehicle components and systems.

NATEF Abbreviation for National Automotive Technicians Education Foundation.

National Automotive Technicians Education Foundation (NATEF) A foundation with a program of certifying secondary and post-secondary automotive and heavy-duty truck training programs.

National Institute for Automotive Service Excellence (ASE) A nonprofit organization that has an established certification program for automotive, heavy-duty truck, auto body repair, engine machine shop technicians, and parts specialists.

Needlenose Pliers Tool with tapered jaws for grasping small parts or for reaching into tight spots. Many needlenose pliers also have cutting edges and a wire stripper.

NIASE Abbreviation for National Institute for Automotive Service Excellence, now abbreviated ASE.

NIOSH Abbreviation for National Institute for Occupation Safety and Health.

NLGI Abbreviation for National Lubricating Grease Institute.

NHTSA Abbreviation for National Highway Traffic Safety Administration.

OEM Abbreviation for original equipment manufacturer.

Ohm A unit of electrical resistance.

Ohm's Law Basic law of electricity stating that in an electrical circuit, current, resistance, and pressure work together in a mathematical relationship.

Open Circuit An electrical circuit whose path has been interrupted or broken either accidentally (a broken wire) or intentionally (a switch turned off).

Oscillation Movement in either fore/aft or side-to-side direction about a pivot point.

OSHA Abbreviation for Occupational Safety and Health Administration.

Out-of-Round Eccentric.

Output Driver Computer output switches in an ECM or ECU. Output drivers are located in the ECU along with the input conditioners, microprocessor, and memory.

Oval Not round; egg-shaped or eccentric.

Overrunning Clutch A clutch mechanism that transmits torque in one direction only.

Parallel Circuit An electrical circuit that provides two or more paths for current flow.

Parts Requisition Form used to order new parts, on which the technician writes the part(s) needed along with the vehicle's VIN.

Payload The weight of the cargo carried by a truck, not including the weight of the body.

PG Propylene Glycol Commonly used antifreeze agent. Should not be mixed with EG.

Pitting Surface irregularities resulting from corrosion.

Polarity The state, either positive or negative, of two electrical poles.

Pounds per Square Inch (psi) A unit of English measure for pressure.

Pressure Force applied to area measured in pounds per square inch (psi) English or kilopascals (kPa) metric.

Pressure Differential The difference in pressure between any two points of a system or a component.

Pressure-Relief Valve Valve located on an air conditioning compressor or pressure vessel that opens if an excessive system pressure is exceeded.

Printed-Circuit Board Electronic circuit board made of thin, nonconductive material onto which conductive metal, such as copper, has been deposited. The metal is then etched by acid, leaving metal lines that form conductive paths for the various circuits on the board. A printed circuit board can hold many complex circuits in a small area.

Programmable Read Only Memory (PROM) A computer memory component that contains program information specific to calibration.

PROM Abbreviation for Programmable Read Only Memory.

psi Abbreviation for pounds per square inch.

RAM Abbreviation for random access memory.

Ram Air Air forced into the engine or passenger compartment by the forward motion of the vehicle.

Random Access Memory (RAM) The memory used during computer operation to store temporary information. The microcomputer can write, read, and erase information from RAM in any order, which is why it is called random.

RCRA Abbreviation for Resource Conservation and Recovery Act.

Reactivity The characteristic of a material that enables it to react violently with air, heat, water, or other materials.

Read-Only Memory (ROM) A type of memory used in microcomputers to store information permanently.

Recall Bulletin A bulletin that pertains to service work or replacement of components in connection with a recall notice.

Reclaim The process used to restore used refrigerants to new product quality. It may include processes available at reprocessing facilities such as distilling and chemical analysis.

Recovery The process of removing refrigerant from a system and storing it in an external container for further processing or reuse.

Recycling The process of cleaning and removing oil, moisture, and acidity from refrigerant, usually at a repair shop, prior to its reuse.

Reference Voltage The voltage supplied to a sensor by the computer, which acts as a baseline voltage; modified by the sensor to act as an input signal. Usually 5 VDC.

Refractometer Refracts light through a liquid to measure specific gravity.

Relay An electric switch that allows a small current to control a much larger one. It consists of a control circuit and a power circuit.

Refrigerant A liquid capable of vaporizing at a low temperature.

Refrigerant Management Center Equipment designed to recover, recycle, and recharge an air conditioning system.

Resistance The opposition to current flow in an electrical circuit.

Revolutions per Minute (rpm) The number of complete turns a member makes in 1 minute.

Right to Know Law A law passed by the federal government and administered by the Occupational Safety and Health Administration (OSHA) that requires any company that uses or produces hazardous chemicals or substances to inform its employees, customers, and vendors of any potential hazards that may exist in the workplace as a result of using the products.

ROM Abbreviation for read-only memory.

Rotation A term used to describe a gear, shaft, or other device that is turning.

rpm Abbreviation for revolutions per minute.

Rotor The rotating member of an assembly.

Runout The deviation or wobble of a shaft or wheel as it rotates. Runout is measured with a dial indicator.

Screw-Pitch Gauge A gauge used to check the threads per inch of a nut or bolt.

Semiconductor Solid state device that can function as either a conductor or an insulator, depending on how it is biased.

Sensor Electronic device used to monitor conditions for input to a computer.

Series Circuit A circuit with only one path for electron flow.

Series/Parallel Circuit A circuit designed so that both series and parallel combinations exist within the same circuit.

Service Bulletin A publication that provides service tips, field repairs, product improvements, and related information of benefit to service personnel.

Service Manual A manual, published by the manufacturer that contains service and repair information for vehicle systems and components.

Short Circuit An undesirable connection between two worn or damaged wires.

Solenoid An electromagnet used to perform mechanical work, made with coil windings wound around an iron tube.

Solid State Device A device that requires little power to operate and generates little heat.

Solid Wire A single-strand conductor.

Solvent A substance that dissolves other substances.

Spade Fuse A term used for blade fuse.

Spalling Surface fatigue that occurs when chips, scales, or flakes of metal break off due to fatigue rather than wear.

Specialty-Service Shop A shop that specializes in areas such as engine rebuilding, transmission/axle overhauling, brake, air conditioning/heating repairs, and electrical/electronic work.

Specific Gravity The scientific measurement of a liquid based on the ratio of the liquid's mass to an equal volume of distilled water.

Spontaneous Combustion A process by which a combustible material ignites by itself.

Static Balance Balance at rest, or still balance.

Stepped Resistor A resistor designed to have two or more fixed values by connecting wires to several taps.

Storage Battery Source of direct current electricity for electrical and electronic systems.

Stranded Wire Wire made up of a number of small solid wires, generally twisted together, to form a single conductor.

Structural Member A primary load-bearing portion of the body structure that affects its over-the-road performance or crash-worthiness.

Swage To reduce or taper.

Switch Device used to control on/off and direct current flow in a circuit. A switch can be under the control of the driver or it can be self-operating.

System-Protection Valve A valve to protect against an accidental loss of pressure, buildup of excess pressure, or back-flow and reverse air flow.

Tachometer An instrument that indicates rotating speeds, sometimes used to indicate crankshaft rpm.

Tapped Resistor A resistor designed to have two or more fixed values, available by connecting wires to several taps.

Tandem A term used to describe an item directly in front of the other and working together.

Throw Number of output circuits of a switch.

Time Guide Reference used for computing compensation payable by the truck manufacturer for repairs or service work to vehicles under warranty, or for other special conditions authorized by the company.

Torque Rotary force.

Toxicity A statement of how poisonous a substance is.

Tractor A motor vehicle that has a fifth wheel and is used for pulling a semitrailer.

Transistor An electronic device produced by joining three sections of semiconductor materials. Used as a switching device.

Tree-Diagnosis Chart A chart used to provide a sequence for what should be inspected or tested when troubleshooting a repair problem.

TTMA Abbreviation for Truck and Trailer Manufacturers Association.

TVW Abbreviation for (1) Total vehicle weight; (2) Towed vehicle weight.

UNEP Abbreviation for United Nations Environment Program. Mandates the eventual phaseout of CFC-based refrigerants.

United Nations Environmental Program (UNEP) A protocol that mandated the eventual phase-out of CFC-based refrigerants.

Vacuum An absence of matter often used to describe any pressure below atmospheric pressure.

Validity List A list of valid bulletins supplied by the manufacturer.

VIN Abbreviation for Vehicle Identification Number.

Viscosity Resistance to fluid sheer. Viscosity describes oil thickness or resistance to flow.

Volt Unit of electromotive force.

Voltage-Generating Sensors These are devices that produce their own input voltage signal.

Watt Measure of electrical power.

Watt's Law Law of electricity used to find the power consumption in an electrical circuit expressed in watts. It states that power equals voltage multiplied by current.

Windings Coil of wire found in a relay or an electrical clutch that provides a magnetic field.

Work The product of a force.

Yield Strength The highest stress a material can stand without permanent deformation or damage, expressed in pounds per square inch (psi).

Zener Diode A variation of the diode, this device functions like a standard diode until a certain voltage is reached. When voltage level reaches this point, the zener diode will allow current to flow in the reverse direction.

Notes

Notes

Notes

Notes